$25—

# Baroclinic
# Processes
# on Continental
# Shelves

## Coastal and Estuarine Sciences

Christopher N. K. Mooers, Series Editor

A series devoted to advancing knowledge of physical, chemical, and biological processes in coastal and estuarine regimes and their relevance to societal concerns.

1. **Coastal Upwelling,** Francis A. Richards (Ed.)
2. **Oceanography of the Southeastern U.S. Continental Shelf,** L. P. Atkinson, D. W. Menzel, and K. A. Bush (Eds.)
3. **Baroclinic Processes on Continental Shelves,** Christopher N. K. Mooers (Ed.)

# Baroclinic Processes on Continental Shelves

Christopher N. K. Mooers
Editor

American Geophysical Union
Washington, D. C.
1986

Baroclinic Processes on Continental Shelves

**Library of Congress Cataloging in Publication Data**

Main entry under title:

Baroclinic processes on continental shelves.
   (Coastal and estuarine sciences ; 3)
   1. Ocean waves—Addresses, essays, lectures.
2. Continental shelf—Addresses, essays, lectures.
I. Mooers, C. N. K. (Christopher N. K.) II. Series.

GC213.B37  1986      551.47      85-31551
ISBN 0-87590-252-9
ISSN 0733-9569

# CONTENTS

# Preface

The AGU Monograph Series on Coastal and Estuarine Regimes provides timely summaries and reviews of major process and regional studies, both observational and theoretical, and of theoretical and numerical models. It grew out of an IAPSO/SCOR/ECOR working group initiative several years ago intended to enhance scientific communications on this topic. The series' authors and editors are drawn from the international community. The ultimate goal is to stimulate bringing the theory, observations, and modeling of coastal and estuarine regimes together on the global scale.

The study of coastal and estuarine regimes is important scientifically because they are where the "oceans meet the continents." In other words, it is through the estuaries and the coastal oceans that materials are exchanged between the oceans and continents. From a geophysical fluid dynamics perspective, estuarine and coastal waters present a rich array of challenging phenomena due to the extreme ranges of density stratification and topographic variation encountered there. Especially in the coastal oceans, the effects of the earth's rotation (Coriolis force) are dominant, too. The coastal oceans are notable for their intense variability, in part due to their extraordinary responsiveness to the passage of atmospheric weather systems. Another great source of variability in addition to river runoff and atmospheric forcing has only recently been appreciated: meandering boundary currents and synoptic/mesoscale eddies of the open ocean impinge upon the continental margins and entrain waters from the coastal oceans. Conversely, some of the eddies of the open ocean may originate from unstable flows in the coastal ocean. Closely associated with the intense physical variability of coastal and estuarine regimes is the rich and varied structure of coastal ecosystems, with their well-known high concentrations and productivity at all trophic levels. How the physical and biological aspects of the coastal and estuarine systems are connected is yet to be fully understood.

The first volume in the series was entitled "Coastal Upwelling," which summarized the state of multidisciplinary knowledge, on an international basis, of the coastal upwelling process as of 1980. The second volume was entitled "Oceanography of the Southeastern U.S. Continental Shelf," which summarized the state of multidisciplinary knowledge on an entire continental shelf regime, presenting evidence for the role of Gulf Stream meanders in driving the coastal ecosystem of interest through eddy fluxes of nutrients from offshore, a striking new finding which reversed conventional wisdom.

The present volume on physical processes in shelf regimes reviews the dynamical topics of coastal trapped waves; shelf break circulation processes, internal tides and waves and near-inertial motions, the coastal boundary layer and inner shelf, estuarine-shelf interactions, coastal and estuarine fronts, and processes which affect density stratification in shelf waters. Hence a significant fraction of the large variety of processes which influence the mass field and circulation variability on a broad range of space and time scales on continental shelves is treated. These processes must be taken into account in designing physical and aphysical sampling programs and also in designing physical and aphysical theoretical and numerical models.

Other forthcoming volumes in the series will treat dynamical and modeling topics, as well as further process and regional studies. Scientists interested in organizing and editing volumes for this series should contact the Series Editor or AGU headquarters for further information.

Christopher N. K. Mooers
Series Editor

# COASTAL TRAPPED WAVES

J. M. Huthnance

Institute of Oceanographic Sciences, Bidston Observatory
Birkenhead, Merseyside L43 7RA, England

L. A. Mysak

Department of Mathematics, University of British Columbia
Vancouver, British Columbia, Canada V6T 1W5

D.-P. Wang

Argonne National Laboratory, Argonne, Illinois 40439

Abstract. We treat waves extending across the continental shelf and/or slope and having periods of the order of 1 pendulum day or longer. Most propagate cyclonically around the ocean unless reversed in trenches or by mean currents; the latter may induce wave growth. Large amplitudes and alongshore topographic changes cause distortion and transfers to other wave modes. We emphasize the waves' role in adjusting oceanic currents to the shelf profile and in propagating wind-driven currents and upwelling along the shelf.

## 1. Introduction

Many shelf seas are dominated by motions which extend across the width of the shelf and fluctuate on time scales of a day or more. Oceanic tides adjust to the shelf topography, contributing large coastal sea level variations and currents, particularly on broad shelves. Atmospheric pressure and (especially) winds over shallow shelf seas can generate strong currents and large changes of sea level: "storm surges" in extremis. To these phenomena, long known from observations of coastal sea level, should be added upwelling, "slope" currents and "poleward undercurrents" ubiquitous on the oceans' eastern shelves, and the shelf response to impinging oceanic eddies and imposed longshore pressure gradients.

Study of the natural waves that travel along or across the continental shelf and slope is a useful preliminary to considering all these shelf motions. We treat here all long waves (horizontal scales comparable with the shelf width) which can occur with periods generally greater than half a pendulum day. Only in passing do we mention "edge" waves, which occur only at shorter periods and are more naturally treated in the context of nearshore and beach processes, for which they appear to be most important. Internal waves and tides, depending on density stratification and also propagating only at shorter periods, are treated separately in this volume. For a historical review, see Mysak [1980].

## 2. Formulation

We consider an incompressible sea, of density $\rho(\underset{\sim}{x}, z, t)$ close to a constant value $\bar{\rho}$, between a gently sloping seafloor $z = -h(\underset{\sim}{x})$ and the free surface $z = \eta(\underset{\sim}{x}, t)$, where $\eta = 0$ for the sea at rest. The Cartesian coordinates $\underset{\sim}{x}, z \equiv (x, y), z$ rotate with a vertical component $\bar{f}/2$. Nearly horizontal motion $\underset{\sim}{u}, w = (u, v), w$ satisfies the following field equations:

Horizontal momentum

$$\partial\underset{\sim}{u}/\partial t + (\underset{\sim}{u}\cdot\nabla\underset{\sim}{u} + w\,\partial\underset{\sim}{u}/\partial z) + f\underset{\sim}{k} \times \underset{\sim}{u}$$
$$\text{sections 6, 7}$$

$$= -\nabla p/\bar{\rho} + (\partial\tau/\bar{\rho}\partial z) \quad (1)$$
$$\text{section 9}$$

Vertical momentum (hydrostatic balance)

$$0 = \partial p/\partial z - \rho g \quad (2)$$

Incompressibility

$$\nabla\cdot\underset{\sim}{u} + \partial w/\partial z = 0 \quad (3)$$

Continuity (sections 5 and 6)

$$\partial\rho/\partial t + (\underset{\sim}{u}\cdot\nabla\rho) + w\,\partial\rho/\partial z = 0 \quad (4)$$
$$\text{section 6}$$

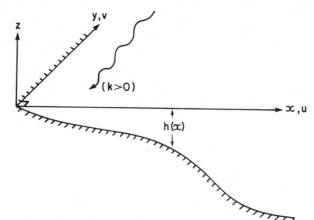

Fig. 1. Straight shelf geometry.

Here $t$ is time, $\nabla \equiv (\partial/\partial x, \partial/\partial y)$, $\underset{\sim}{k}$ is the upward vertical unit vector, $\underset{\sim}{\tau}$ is the horizontal stress vector, and $g$ is downward gravitational acceleration. Lateral friction and density diffusion are neglected as secondary to the presence of the waves to be considered.

Boundary conditions are as follows:

At the seafloor $z = -h$

$$w = -\underset{\sim}{u}\cdot\nabla h \qquad (5)$$

At the sea surface $z = \eta$

$$p = 0 \qquad (6)$$

or $p$ is atmospheric pressure $p^a$, (section 9)

$$w = \partial\eta/\partial t + (\underset{\sim}{u}\cdot\nabla\eta) \qquad (7)$$

sections 6, 7

$\underset{\sim}{\tau}$ is specified (=0 for free waves) $\qquad (8)$

At the coast $x = 0$

$$u = 0 \text{ or } u \text{ is finite if } h \to 0 \qquad (9)$$

Far from the coast, $x \to \infty$

$$u \to 0 \qquad (10)$$

(or $u$ is specified, section 9).

(The density $\rho$ is simply advected by (4) and requires no separate boundary condition.)

Most of the following treats simplified equations, the bracketed terms being used only in the sections indicated. Furthermore, for flow independent of $z$, integration of (3) through depth, and application of (5), (7) yields

$$\partial\eta/\partial t + \nabla\cdot(h\underset{\sim}{u}) + [\nabla\cdot(\eta\underset{\sim}{u})] = 0 \qquad (11)$$

sections 6, 7

which also holds for the depth-averaged flow $\underset{\sim}{u}$ and replaces (3), (5), and (7). Moreover if $\rho$ is uniform, we may substitute in (1)

$$\nabla p/\overline{\rho} = g\nabla\eta + (\nabla p^a/\overline{\rho}) \qquad (12)$$

section 9

from (2) and (6), which are no longer required.

### 3. Waves on a Straight Unstratified Shelf

We take $x$ offshore and $y$ alongshore as in Figure 1; the depth is $h(x)$ independent of $y$. We seek solutions

$$\{u(x), v(x), \eta(x)\} \exp(iky + i\sigma t)$$

independent of $z$. By convention, $\sigma > 0$. Then $k > 0$ corresponds to propagation in $-y$, with the coast on the right. Equation (1) with (12) gives $u$ and $v$ in terms of $\eta$, which satisfies

$$(h\eta')' + K\eta = 0$$
$$K \equiv kfh'/\sigma + (\sigma^2 - f^2)/g - k^2h \qquad (13)$$

by (11) if $f$ is uniform: primes denote $\partial/\partial x$. The boundary conditions (9) and (10) imply

$$h(\sigma\eta' + fk\eta) \to 0 \qquad x \to 0$$
$$\eta \to 0 \qquad x \to \infty \qquad (14)$$

Equations (13) and (14) may also be derived as a low-stratification limit of (1)-(10) [Huthnance, 1978a].

Longshore variations of $f$ are considered in section 8. Cross-shelf variations of $f$ are important for zonally traveling waves near the equator whenever $f'/f \gtrsim h'/h$ [Mysak, 1978a].

A generally applicable formulation for low frequencies ($\sigma \lesssim f$) is to neglect $\partial\eta/\partial t$ in (11), implicitly assuming

$$\sigma^2 L^2/gh \ll 1 \qquad f^2 L^2/gh \ll 1$$

where $L$ is the cross-shelf scale. Hence there is a stream function $\Psi$: $hu = -ik\Psi$, $hv = \Psi'$ satisfying

$$(\Psi'/h)' + \hat{K}\Psi = 0 \qquad \hat{K} \equiv (-k/\sigma)(f/h)' - k^2/h \qquad (15)$$

with boundary conditions

$$\Psi \to 0 \qquad x \to 0, \infty \qquad (16)$$

Free wave modes are represented by eigensolutions of (13) and (14). Successive modes with more offshore nodes correspond to more positive $K$ and arise in two ways, for a given wave number $k$ along the shelf.

1. The term $(\sigma^2 - f^2)/g - k^2h$ increases with $\sigma$. The term $(\sigma^2 - f^2)/g$ represents the familiar surface gravity wave mechanism, modified by the rotation; note its absence from (15). Usual term-

inology is "Kelvin wave" for the mode with no nodes offshore and $kf > 0$, and "edge wave" otherwise.

2. The term $kfh'/\sigma$ increases with decreasing $\sigma$, if h is monotonic and $kf$ is positive corresponding to cyclonic propagation around deep water. The term $kfh'/\sigma$ (or $-k/\sigma$ $(f/h)'$ in (15)) represents the following potential vorticity restoring mechanism [Longuet-Higgins, 1972]. If fluid is displaced into shallower water, it spreads laterally to conserve volume and therefore spins more slowly in total; i.e., it acquires anticyclonic relative vorticity. Along the slope in the sense of cyclonic propagation around deep water, the resulting upslope velocity implies an upslope displacement in time. Hence the disturbance propagates along the slope and is restored by the subsequent velocity field. We refer to these modes simply as "shelf waves."

The mode forms and frequencies are known for a variety of analytic models including level, uniformly sloping, and exponential concave and convex shelves adjacent to an ocean of uniform depth. The earliest model [Reid, 1958] treated an unbounded uniform slope. More details and original references are given by LeBlond and Mysak [1978, pp. 219-240]; see also Clarke [1974].

For any monotonic depth profile h(x) it can be shown that (1) all modes with $\sigma < |f|$ propagate cyclonically about the deep sea (i.e., $kf > 0$); (2) the shelf waves with 1, 2, . . . offshore nodes have frequencies $|f| > \sigma_1(k) > \sigma_2(k) > $ . . . defined for all k (subject to $kf > 0$); (3) the Kelvin wave frequency $\sigma_0(k) > \sigma_1(k)$ is likewise defined for all k ($kf > 0$); (4) the edgewaves with 0 (if $kf < 0$), 1, 2, . . . offshore nodes have increasing frequencies $\sigma > |f|$ but are subject to a low-wave number/low-frequency cutoff where the dispersion curves break the trapping criterion [Huthnance, 1975]

$$0 < K(\infty) \equiv (\sigma^2 - f^2)/g - k^2h(\infty)$$

Numerical solutions for the waveforms, and dispersion relations $\sigma(k)$, are easily found for any depth profile h(x) following Caldwell et al. [1972].

Laboratory experiments for shelf waves gave good agreement with theory [Caldwell et al., 1972].

Illustrative dispersion curves and waveforms are shown in Figure 2 obtained numerically for h(x) (shown in Figure 2c) representing the fairly broad shelf and steep slope west of Scotland. Besides properties 1-4, other features are typical. For shelf waves, bounded $h'/h$ would ensure the maximum in $\sigma(k)$; here the group velocity $\partial\sigma/\partial k$ of energy propagation (in −y) reverses through zero. As $k \to \infty$, $\sigma_n \to f/(2n + 1)$ for the n-node shelf wave which becomes concentrated over the "beach" near $x = 0$. As $\sigma, k \to 0$, the shelf wave and Kelvin wave speeds $\sigma/k$ approach constant (maximum) values, and $u/\sigma$, v, $\eta$ approach constant

forms, so that cross-slope velocities tend to zero. Variables v, $\eta$, and $\eta'$ are in phase or antiphase, v and $\eta'$ being near "geostrophic" balance $fv = g\eta'$ by (1); u is 90° out of phase, giving cyclonically polarized Kelvin wave currents on the shelf but anticyclonically polarized currents for the first mode shelf wave. The Kelvin wave speed decreases from near $(gH_3)^{1/2}$ to near $(gH_2)^{1/2}$ for increasing $k = O(x_2^{-1})$, or $\sigma = O(gH_2)^{1/2}/x_2$, i.e., as 1/4 gravity wavelength decreases to the shelf width $x_2$ (Figure 2c) and the Kelvin wave "climbs" onto the shelf.

Shelf wave forms depend on the shape rather than the horizontal scale L of the depth profile; phase speeds scale as fL and (u, v, $\eta$) scale as $(\sigma U/f, U, fUL/g)$, where U is typically 0.1 m s$^{-1}$ (see sections 9 and 10). Kelvin wave forms depend more on the depth scale; usually, the phase speed is just under $(gh(\infty))^{1/2}$, and (u, v, $\eta$) scale as $(\sigma Z/hL, (g/h)^{1/2}Z, Z)$ where Z is typically 0.1-1 m (sections 9 and 10). However, quantitative results depend on accurate profile modeling, and we advocate numerical calculations for real shelves.

The equatorial case (15) has been reviewed by Mysak [1980]. As expected from the similarity to (13), shelf waves are not greatly affected if $(f/h)'$ is one-signed, which is more or less assured if the sea does not straddle the equator. However, if the sea does straddle the equator, waves propagating in the same sense as shelf waves in the coastal hemisphere are subject to a low-frequency cutoff: for trapping, $0 > K(\infty)$, i.e., $\sigma > -f'/k$. Moreover, there is an additional infinite set of wave modes propagating in the opposite sense along the lower slope if the sloping shelf straddles the equator. All frequencies are low by virtue of small f.

4. Other Geometry

Rectangular-basin modes have been reviewed by LeBlond and Mysak [1978]. Rotating basins are more readily treated if circular; (13) is replaced by

$$(hr\eta')' + M\eta = 0$$
$$M \equiv mfh'/\sigma + r(\sigma^2 - f^2)/g - m^2h/r \qquad (17)$$

where r is the radial coordinate, primes denote $\partial/\partial r$, $h = h(r)$, and m is the azimuthal wave number; m is quantized: $m = 0, \pm1, \pm2, \ldots$. Otherwise, (17) is very similar to (13). Analyses have been performed for $h \propto 1 - r^2/a^2$ by Lamb [1932, p. 326], subject to suitable interpretation [Miles and Ball, 1963], and for $h \propto (1 - r^2/a^2)^2$ by Saint-Guily [1972]. Properties 1-4 of section 3 persist, neglecting the node at $r = 0$ and replacing continuous k by quantized −m.

For circular islands, however, if $\sigma > |f|$, then $M > 0$ as $r \to \infty$, and trapping cannot be perfect, unless (improbably!) $h/r^2 \not\to 0$. LeBlond and Mysak [1978, pp. 246-262] have reviewed this case: for

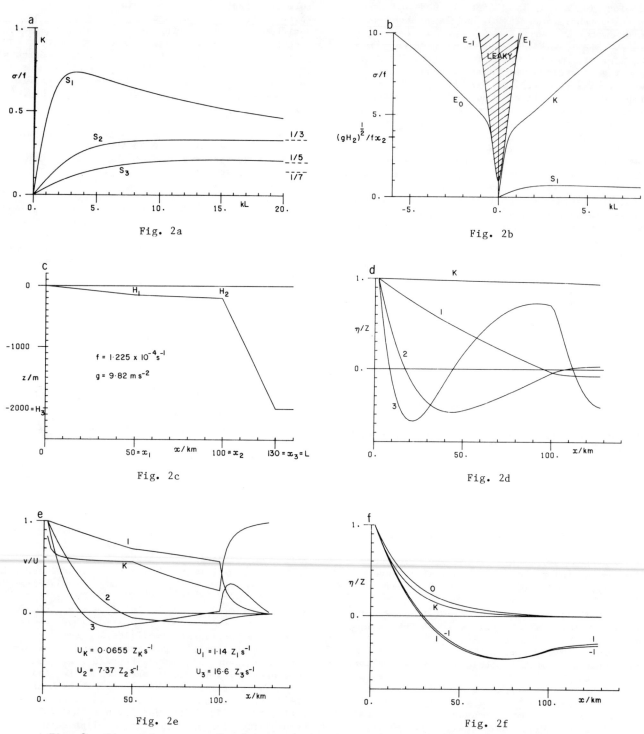

Fig. 2a

Fig. 2b

Fig. 2c

Fig. 2d

Fig. 2e

Fig. 2f

Fig. 2. Dispersion curves for (a) n-node shelf waves $S_n$, (b) Kelvin wave K and $|n|$-node edge waves $E_n$, for (c) depth profile. Phase speeds $\sigma/k$ at the origin are 127, 8.4, 1.45, and 0.71 m s$^{-1}$ for K, $S_1$, $S_2$, and $S_3$. (d) Kelvin wave (K) and shelf wave (1, 2, 3) elevations and (e) longshore currents v for $\sigma/f = 0.01$. (f) Kelvin wave (K) and edge wave (-1, 0, 1) elevations for $\sigma/f = 10$. Positive v is forward under the wave crest ($\eta > 0$).

islands with shelves, there may be frequencies $\sigma(m)$ at which waves almost trapped on the shelf lose energy by radiation to the deep sea only very slowly (see also Lozano and Meyer [1976]). Mysak [1980] reviews the case $\sigma < |f|$; shelf waves exist qualitatively as in section 3 (but quantized m!), i.e., propagating anticyclonically around the island, provided $h' > 0$ for some r. There is always at least one "Kelvin" wave ($m = -f/|f|$) in the same sense, even if $h' = 0$ [Longuet-Higgins, 1969]; further Kelvin waves, e.g., $m = -nf/|f|$, require a very large island radius

$$a > (n(n - 1)gh)^{1/2}/f \quad \text{if } h' = 0$$

For a broad straight shelf, the continental slope may be regarded as a scarp, neglecting the distant coast. LeBlond and Mysak [1978, pp. 178-184, 214-217] review this context. Only waves with $\sigma < |f|$ are trapped. They are in effect a Kelvin wave and a complete set of shelf waves over the slope, with properties as in section 3, propagating cyclonically relative to the deep water. However, the Kelvin wave is now "double," decaying exponentially on both sides.

If the scarp straddles the equator, the waves have low frequency but may propagate in either sense (on the appropriate side) subject to $\sigma > -f'/k$ [Green and Buchwald, 1979].

A curved scarp, closing on itself to form a seamount, retains the same set of trapped waves, but the long-slope wave number is quantized so that there close comparison with the shelf around an island. Circular and elliptic depth contours have been studied analytically [Rhines, 1969a, b; Huthnance, 1974].

A ridge comprises two scarps back-to-back. Each retains its set of "shelf" waves ($\sigma < |f|$) propagating in the appropriate sense, and there is a double Kelvin wave associated with any net depth change [Brink, 1983]. "Edge" waves also propagate in both senses along the ridge with $\sigma/|f|$ large enough to cause $K > 0$ in (13) (over the ridge; $K < 0$ for trapping on either side) [Buchwald, 1969].

Some shelves do not have monotonic depth. Louis [1978] considered a level shelf with a shelf-edge ridge. The $\sigma < |f|$ ridge waves appropriate to the reversed slope persist, hardly affected by the coast. Mysak et al. [1979] considered a coastal trench. Again, the sets of waves appropriate to both slopes are present, slightly modified by more rapid decay over the opposing slope.

All cases $h = h(x)$ and $h = h(r)$ may be easily treated numerically as indicated in section 3.

### 5. Stratification

We consider a straight shelf (Figure 1) with a rest state $\rho = \rho_0(z)$, $p = p_0(z)$ satisfying (2), let

$$N^2 = -g\, d\rho_0/\bar{\rho}\, dz \qquad \rho = \rho' + \rho_0 \qquad p = p' + p_0$$

and seek solutions

$$\left\{ u'(x,z),\ v'(x,z),\ w'(x,z),\ \rho'(x,z),\ p'(x,z) \right\}$$
$$\cdot \exp(iky + i\sigma t)$$

Dropping primes, (2), (4), and (1) give $\rho$, $w$, $u$ and $v$, respectively, in terms of $p$, which satisfies

$$\partial^2 p/\partial x^2 + (f^2 - \sigma^2)\, \partial/\partial z(N^{-2}\, \partial p/\partial z) - k^2 p = 0 \quad (18)$$

by (3). The boundary conditions (5), (9), (6), (7), and (10) imply

$$\partial h/\partial x(\partial p/\partial x + fkp/\sigma) + (f^2 - \sigma^2)N^{-2}\, \partial p/\partial z = 0$$
$$z = -h \quad (19)$$

$$\partial p/\partial z + N^2 p/g = 0 \qquad z = 0 \quad (20)$$

$$p \to 0 \qquad x \to \infty \quad (21)$$

where (19) is interpreted as $\partial p/\partial x + fkp/\sigma = 0$ at any vertical scarps or if $h \neq 0$ at the coast. A flat bottom $h = const$ with $N^2 = const$, $N^2 h/g \ll 1$ ($\rho - \bar{\rho} \ll \bar{\rho}$) admits the simplest Kelvin wave and ($n \geqslant 1$) "internal Kelvin wave" solutions

$$\cos(N(z + h)/c_n) \exp(i\sigma t + i\sigma y/c_n - fx/c_n) \quad (22)$$

$$c_n = (gh)^{1/2} \qquad n = 0$$
$$c_n = Nh/n\pi \qquad n = 1, 2, \ldots$$

approximately, at all frequencies $\sigma$. Similar solutions, with vertical structure distributed roughly as N, exist for any $N(z) > 0$. Propagation is in the same sense as for shelf waves, cyclonic about the deep sea, or anticyclonic around a cylindrical island [Hogg, 1980], but depends only on density stratification.

For a sloping bottom $h(x)$ with offshore scale L (e.g., shelf width) and depth scale H, the parameter $S \equiv N^2 H^2/f^2 L^2$ indicates the importance of stratification [Huthnance, 1978a]. For small S the solutions of (18)-(21) in $\sigma < |f|$ are depth-independent Kelvin and shelf waves. As S increases, the wave speeds increase, and nodes of $u$, $v$, $p$ in $(x, z)$ tilt outward from the vertical toward horizontal. Quite moderate S may imply $\sigma$ monotonic increasing in k, contrasting with the maximum in $\sigma(k)$ common among unstratified modes [Chapman, 1982b, 1983]. For large S the modes in $\sigma < |f|$ become internal Kelvin-like waves (22) with x replaced by $x - h^{-1}(z)$. $S \to \infty$ means shelf width L much less than "internal deformation scale" $NH/f$; the slope is "seen" only as a coastal wall.

For $\sigma > |f|$, trapping is imperfect over a shelf $h(x)$ if $S > 0$. However, there may be frequencies σ(k) at which almost trapped waves, corresponding to edge waves, radiate energy very slowly; Mysak [1968] and Chapman [1982a] consider two-layer and continuously stratified examples, respectively.

Bottom-trapped waves have motion everywhere parallel to a plane sloping seafloor (in uniform $N^2$) and decay exponentially away from the seafloor if the sea surface is neglected [Rhines, 1970]. They may propagate for $\sigma \gtrless |f|$ and up or down the slope, but always with a component along the slope in the sense of shelf waves and internal Kelvin waves. In Rhines' [1970] description, they are not confined laterally in any way, but in $\sigma < |f|$, solutions of (18)-(21) tend for large k to bottom-trapped waves confined near the seafloor maximum of $Nh' < |f|$ and have frequency $Nh'$ [Huthnance, 1978a]. Chapman and Hendershott [1982] and Chapman [1983] treat in more detail some examples with uniform or exponential stratification and various depth profiles. If max $Nh' > |f|$, then σ increases to $|f|$ as k increases, a qualitative difference from unstratified behavior [Chapman, 1982b]. We speculate that as k increases further (to infinity), the waves still tend to (slightly leaky) bottom-trapped waves with $\sigma \sim Nh' > |f|$; Rhines' [1970] solution is still valid in $\sigma > |f|$.

Idealizations with two-layer stratification and vertical scarps cannot even approximate a maximum of $Nh'$ and hence are useful only for medium or long waves (small k). Modes with significant vertical structure tend to be represented as non-dispersive Kelvin waves, owing to (1) a well-defined upper layer depth, (2) scarp width much less than internal deformation radius, and (3) a lower layer (offshore from the scarp) of constant depth or much deeper than the upper layer. The combination 1-3 is of limited applicability; in particular, property 2 generally holds only at low latitudes. Two-layer stratification also erroneously predicts zero oceanward energy leakage in $\sigma > |f|$ for all sufficiently large k. The small-σ leaky modes for a uniformly stratified wedge [Ou, 1980] suggest that oceanward energy leakage via bottom-trapped waves occurs for sufficiently small σ ($< |f|$!) unless $Nh' \to 0$ as $x \to \infty$ along the seafloor.

Stratification and topography models must therefore be chosen carefully. Unfortunately, analysis is rarely possible except for two-layer models.

These include models where the interface meets a vertical wall at the coast above a sloping bottom; Allen [1975] considers weak stratification (small $gh\Delta\rho/\bar{\rho}f^2L^2$ replacing S) so that internal Kelvin and shelf waves are distinguishable, and Wang [1975] treats an exponential depth profile numerically. Kawabe [1982] considered two layers adjacent to upper and lower slopes, finding shelf waves proper to each slope segment and an internal Kelvin-type wave near the interface/bottom intersection; the internal Kelvin-type wave becomes fastest for large wave numbers. Double-Kelvin waves in a two-layer fluid over a scarp were found

by Rhines [1977], for a low scarp and negligible surface displacements, and by Bondok [1980]. Other two-layer models are reviewed by Mysak [1980].

Numerical calculations, less straightforward than in the unstratified case, have been described by Wang and Mooers [1976], Huthnance [1978a], and Brink [1982b]. A variational form of (18)-(21) gives a second-order estimate of the wave speed [Huthnance, 1978a] which will then be accurate in conjunction with poorer wave forms [Brink, 1982b]. The waveforms may be improved using the orthogonality relation

$$0 = \int_{z=-h(x)} p_m p_n \, dz = \int_{z=-h(x)} p_m p_n h' \, dx$$

between different modes $p_m, p_n$ given k, $(f^2 - \sigma^2)/N^2$, and $(f^2 - \sigma^2)/g$ [Huthnance, 1978a].

Illustrative waveforms are shown in Figure 3 for the same depth profile, representing the west Scottish shelf, and representative seasonal stratification over the shelf and permanent stratification below. Like many shelves, this one is much wider than the internal deformation radius, i.e., S is small. Any one mode then becomes depth-independent as $S \to 0$. However, a few modes (of ever higher order as $S \to 0$) retain important vertical structure, as internal mode speeds compare with higher-order shelf wave speeds.

## 6. Mean Flows

Shelf waves with phase speeds of a few meters per second or less may be significantly affected by (in particular) western boundary currents of comparable speed and/or O(f) vorticity.

The purely advective effect of a uniform mean current V according to linearized theory is essentially trivial, although able to reverse the propagation of slower (higher mode or short wavelength) shelf waves. Small modifications due to small shears $\partial V/\partial x \ll |f|$ or $\partial V/\partial z \gg \sigma/kh$ may be derived by a perturbation analysis; Grimshaw [1983] has treated surface and internal Kelvin waves.

Shear $V'(x)$ modifies the background potential vorticity to $P(x) = (f + V')/h$; replacing (15) after writing $hu = (-ik\Psi, \Psi' + hV)$ and retaining $\underset{\sim}{u} \cdot \nabla \underset{\sim}{u}$ in (1) we have

$$(V + \sigma/k)[(\Psi'/h)' - k^2\Psi/h] - P'\Psi = 0 \qquad (23)$$

neglecting $\Psi^2$. Hence the shelf waves are at least modified for $P' \neq (f/h)'$ as well as advected [Brooks and Mooers, 1977a]. Allowing for horizontal divergence, unlike (23), Kenyon [1972] found a "forbidden" range

$$f^2 - (V')^2 < \sigma^2 < f^2$$

for trapped waves on uniform h', V'. This unusual deterrent to Kelvin waves is probably due to the

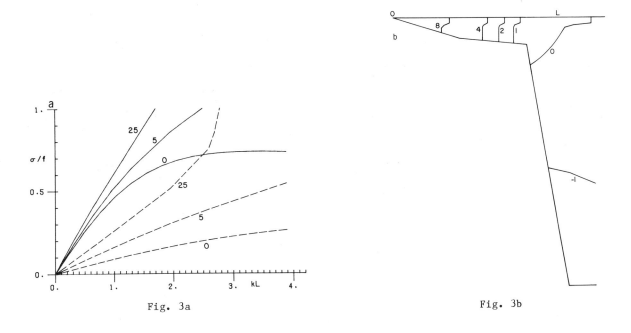

Fig. 3a

Fig. 3b

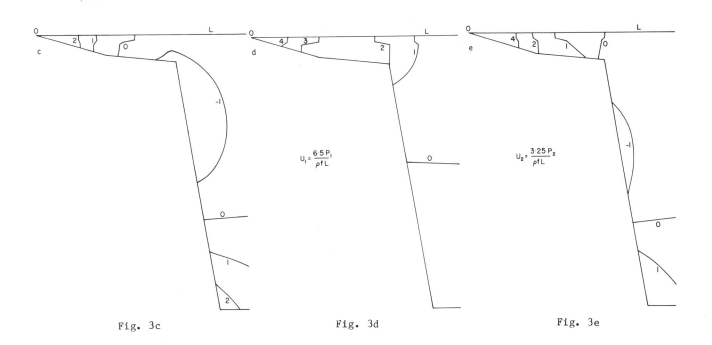

Fig. 3c

$$U_1 = \frac{6 \cdot 5 \, P_1}{\rho f L}$$

Fig. 3d

$$U_2 = \frac{3 \cdot 25 \, P_2}{\rho f L}$$

Fig. 3e

Fig. 3. (a) Dispersion curves for waves on shelf of Figure 2c and $d\rho_0/dz = 0.025F$ kg m$^{-4}$ ($-60 < z < -40$ m: seasonal thermocline), $d\rho_0/dz = 0.25 \times 10^{-3}F$ kg m$^{-4}$ (elsewhere). $F = 1$ represents the west Scottish shelf; F values label curves. ($F \neq 1$ also represents a shelf with the same stratification but relative depth $F^{1/2}$ or relative width $F^{-1/2}$ or relative f value $F^{-1/2}$.) Solid curves represent the first mode, and dashed curves the second mode. (b) First-mode and (c) second-mode pressure fields p/P for $F = 25$, $k = 10^{-4}$. Corresponding longshore velocity v/U for (d) the first mode and (e) the second mode.

infinite offshore extent of the shear. Other models with concave and convex depth profiles are reviewed by Mysak [1980]. For small V and monotonic depth, fP' < 0. If fP' > 0 for some range of x, due to (say) h' < 0 [Mysak, 1982] or V'' > 0 at the outer edge of a coastal jet [Niiler and Mysak, 1971], additional trench or shear waves propagate in the opposite sense (cf. section 4).

"Barotropic" instability is possible if V is strong enough to reverse a shelf wave and give $\sigma(k)$ matching a trench or shear wave. Necessary conditions for instability are $P'(x_s) = 0$ (some $x_s$) and $P'(V(x_s) - V) > 0$ for some x [Collings and Grimshaw, 1980], and the growth rate is bounded by max $|V'/2|$. Collings and Grimshaw [1980] and Hall [1980] also give semicircular bounds (in the complex plane) on the unstable wave speeds. Generally, one expects f/h or depth contours to guide and therefore to stabilize currents according to

$$(\partial/\partial t + \underset{\sim}{u} \cdot \nabla) \left[\frac{f + \underset{\sim}{k} \cdot \nabla \underset{\sim}{u}}{h + \eta}\right] = 0$$

from (1) and (11). Experimentally, Kimura [1976] found depth contours to be stabilizing for a coastal jet opposing the associated shelf waves, because then $V + \sigma/k$ and $P'$ tend to be one-signed in (23). However, there are examples where slopes h'(x) can destabilize an otherwise stable current V(x) [Collings and Grimshaw, 1980]. Growth times may be just a few days.

For nonuniform V, (23) has an additional continuum of solutions for given k, one for each value of phase speed in the range of V. At the "critical layer" $x = x_c$ (say) where $V + \sigma/k = 0$, the solution is bounded but has singular derivatives [McKee, 1979].

In a stratified context, all the above may still apply. Additionally, vertical shear $\partial V/\partial z$ in a mean longshore current is associated with horizontal density gradients: $f \, \partial V/\partial z = -g/\bar{\rho} \, \partial \rho/\partial x$ from (1) and (2).

Two-layer models represent $\partial \rho/\partial x$ by a sloping interface; the varying layer depths are another source of gradients $\partial P/\partial x$ in each layer, giving scope for more waveforms [Iida, 1970; Lacombe, 1982], intersecting dispersion curves, and associated "hybrid" instabilities. "Baroclinic" instability, associated with $\partial V/\partial z$ and extracting the sloping interface's gravitational potential energy, may occur even when V and total depth h are independent of x and do not contribute to $\partial P/\partial x$. More generally, Pedlosky [1964] showed that slow flows over a sloping bottom (relative depth variations in both layers being small) are unstable only if $\partial P/\partial x$ has both signs in the system. Hence in a basic model [Mysak, 1977] with uniform flows $V_1$, $V_2$ along a channel with vertical walls, a uniform cross-channel bottom slope gives stability if it exceeds the interface slope and has the same sense. However, a smaller bottom slope may destabilize long waves, and an opposing slope destabilizes short waves [Mechoso, 1980]. More complex topography is treated by Mechoso and Sinton [1981] and Mysak et al. [1981a]. Bane [1980] numerically finds the stable modes for a shelf with a surface layer jet above the slope. Mysak [1980] reviews other models. However, the same cautions as in section 5 apply; there is a suspicion that two-layer models unduly segregate internal Kelvin and shelf wave types (for example), exaggerating the multiplicity of wave forms and scope for instability.

Laboratory experiments by Griffiths and Linden [1981, 1982] explore both barotropic and baroclinic instability in the simplest context of two layers and uniform total depth.

In continuous stratification the equivalent potential vorticity gradient [LeBlond and Mysak, 1978, p. 423]

$$\partial f/\partial x + \partial^2 V/\partial x^2 + f^2 \, \partial(N^{-2} \, \partial V/\partial z)/\partial z$$

may support waves in the interior. Luther and Bane [1979] found that density contours rising coastward in association with a shelf edge surface jet modified the fastest shelf wave to a "frontal trapped" form inshore of the surface jet. The bottom slope also supports waves. Over a uniform bottom slope, additional bottom features can couple and destabilize the interior and bottom modes [deSzoeke, 1976]. Pedlosky [1980] gives an example of a bottom slope destabilizing an otherwise stable flow V(x, z). However, Wright [1980], extending Mysak's [1977] model with an intermediate uniformly stratified layer, found that bottom-intensified modes were stabilized by the bottom slope.

## 7. Nonlinear Effects

The terms $\underset{\sim}{u} \cdot \nabla \underset{\sim}{u}$ and $\nabla \cdot (\eta \underset{\sim}{u})$ in the equations of motion (1) and (11) generate harmonics and nonzero time averages from a single sinusoidal form $(\underset{\sim}{u}, \eta)$, and sum and difference frequencies from a pair of sinusoids. Changes of waveform, mean current generation, and wave-wave interaction may result.

Nonlinear Kelvin waves in a flat-bottomed sea of total depth h are nondispersive, having speed $(gh)^{1/2}$ for all wavelengths. Indeed, each individual part of the waveform moves with the local speed $(gh)^{1/2}$, so that crests gain on troughs and wave fronts steepen [Bennett, 1973]. Similarly, for internal Kelvin waves on the interface between a surface layer (depth h, density $\rho$) and a deep lower layer of density $\rho + \Delta\rho$ ($\Delta\rho \ll \rho$), the local speed is $(gh \, \Delta\rho/\rho)^{1/2}$ everywhere, and troughs gain on crests [Bennett 1973; Yamagata, 1980]. The same is true along a slowly curving coastline [Clarke, 1977a].

Dispersion limits this nonlinear steepening; the associated shorter wavelengths propagate more slowly in the presence of topography, for Kelvin waves and for any one shelf wave mode; the steepening is "left behind." For small nonlinearity (small amplitudes) and small dispersion (long waves), permanent-form $\text{sech}^2 (ky + \sigma t)$ solutions exist, balancing steepening and dispersion [Smith,

1972; Grimshaw, 1977a; Odulo and Pelinovskiy, 1978].

Steepening may be important for internal Kelvin waves [Yamagata, 1980] but it is too slow to be important for typical shelf wave amplitudes (Grimshaw estimated a 73-day time scale off east Australia).

A particular solution for nonlinear edge waves [Mollo-Christensen, 1979] involves a uniform mean current V fundamentally in the waveform and dispersion relation, contrasting with linear theory (section 6).

Mean currents generated by waves in stratified seas over large topography have been treated by Huthnance [1981] for $\mu \ll \sigma$, f, and by Denbo and Allen [1983] for $\sigma$, $\mu \ll$ f and parallel depth contours; $\mu$ is a frictional decay rate or wave growth rate and must be nonzero to determine the otherwise arbitrary mean flow along f/h contours [Moore, 1970]. (Generation is through the frictional contributions to time averages of $\underset{\sim}{u} \cdot \nabla \underset{\sim}{u}$ in (1) and $\nabla \cdot (\eta \underset{\sim}{u})$ in (11). The mean currents of $O(h^{-1}h'f\xi^2)$ are typically confined close to the coast, forced in either sense by wind-driven shelf waves, or close to the shelf break, where they are in the direction of shelf wave propagation, cyclonic about the deep sea. ($\xi$ is onshore-offshore excursion, of scale $O(U/f)$ (shelf waves) or $O(ZL/h)$ (Kelvin wave) by section 3). Typical values of U = 0.1 m s$^{-1}$, Z = 1 m, h = 100 m, and L = 100 km give mean currents of 0.01 m s$^{-1}$ and 0.1 m s$^{-1}$ for $h^{-1}h'$ = (10 km)$^{-1}$ and (1 km)$^{-1}$, respectively. Ou and Bennett [1979] consider mean currents generated by internal Kelvin waves in a two-layer model with interfacial and bottom friction; a flat bottom was treated analytically, but matching observations required bottom topography, which was treated numerically. The mean currents are $O(U^2(gh \Delta\rho/\rho)^{-1/2})$, or 0.01 m s$^{-1}$ for U = 0.1 m s$^{-1}$, $\Delta\rho/\rho = 10^{-3}$, and h = 100 m, along the coast in the sense of the internal Kelvin wave propagation.

Interactions between shelf waves have been considered by Hsieh and Mysak [1980], formally for any depth profile and explicitly for exponential h(x). For many pairs $\sigma_1$, $k_1$ and $\sigma_2$, $k_2$ $\sigma_3 = \sigma_1 \pm \sigma_2$ and $k_3 = k_1 \pm k_2$ also satisfy the dispersion relation for some shelf wave mode, which initially responds resonantly to the forcing by the nonlinear terms. The energy is drawn from waves 1 and 2. Since $\sigma_1 = \sigma_3 \mp \sigma_2$, etc., all three waves are on an equal footing, changing in proportion to the amplitudes of the other two. The evolutionary time scale is $O(U/L) \sim$ 12 days with the typical values above. Similar behavior occurs for wave triads in a uniformly stratified channel with small cross-channel bottom slope [Mysak, 1978b]. For Mysak's [1977] two-layer channel with small cross-channel bottom slope, waves 2 and 3 of the triad may be the same, with $k_1 = 2k_2$, $\sigma_1 \sim 2\sigma_2$. Then just the two waves interact, and prior presence of ($\sigma_1$, $k_1$) destabilizes its subharmonic ($\sigma_2$, $k_2$), provided either the bottom slope or the current is not uniform, i.e., the potential vorticity gradient varies across the channel [Hukuda, 1982].

Slow modulations in the amplitude of a wave ($\sigma_1$, $k_1$) may be regarded as being due to the presence of another wave ($\sigma_2$, $k_2$) with $\sigma_2 \sim \sigma_1$, $k_2 \sim k_1$; then nonlinearities generate harmonics ($2\sigma_1$, $2k_1$) and differences ($\sigma_1 - \sigma_2$, $k_1 - k_2$) $= (k_1 - k_2)(\partial\sigma/\partial k, 1)$. Grimshaw [1977b] found that the original shelf wave (mode n, say) is stable to these effects, i.e., to slow modulations, if it is long ($k_1L \ll 1$); mode n becomes unstable when $k_1$ increases through

$$\partial\sigma/\partial k|_{k_1} = \sigma/k|_{k=0} \qquad \text{for mode n + 1}$$

is stable again when $k_1$ increases through

$$\partial\sigma/\partial k|_{k_1} = \sigma/k|_{k=0} \qquad \text{for mode n + 2}$$

and so forth.

If wind forces a motion ($\sigma_2$, $k_2$) in the presence of a shelf wave ($\sigma_1$, $k_1$), then the nonlinear terms generate a signal ($\sigma_3$, $k_3$), where

$$\sigma_3 = \sigma_2 - \sigma_1 \qquad k_3 = k_2 - k_1$$

in proportion to the shelf wave amplitude. Barton [1977] considered the exponential shelf wave growth when $\sigma_2 \sim 2\sigma_1$, $k_2 \sim 2k_1$ so that the wave ($\sigma_3$, $k_3$) adds to the original shelf wave. Significant growth in 3 days along the Australian coast is possible. Barton and Buchwald [1977] considered periodic wind forcing $\sigma_2$ with a spectrum of wave numbers $k_2$ over an exponential scarp; a large response occurs if (exceptionally) $\sigma_3$ corresponds to zero $\partial\sigma/\partial k_3$ for some mode, so that a range of wave numbers $k_2$ contribute to the resonant forcing of that mode.

### 8. Longshore Variations

Previous sections have assumed the stratification, depth profile, f, etc., to be uniform alongshore. Apart from a few very simple geometries, analysis of effects of longshore variations is limited to slow, small, random small or abrupt changes.

#### Slow Variations

If changes in the waveguide properties over one wavelength are small, then individual wave modes conserve longshore energy flux, even when slightly nonlinear [Odulo and Pelinovskiy, 1978]. Local waveforms are appropriate to a uniform shelf.

As a consequence, Kelvin wave amplitudes increase as $f^{1/2}$ and are confined closer to the coast at higher latitudes [Miles, 1972]. Their amplitudes also increase, as $h^{-1/2}$, in decreasing depth [Miles, 1973]. Kelvin and internal Kelvin wave speeds are reduced around capes [Miles, 1972; Clarke, 1977a]; Kelvin wave speeds are reduced by a (narrow) shelf and are affected by the earth's curvature along nonmeridional boundaries [Miles, 1972].

Grimshaw [1977c] studied shelf waves in the context of slowly varying f, shelf topography, and coastline orientation. Conservation of energy flux implies a large amplitude increase if waves of frequency $\sigma$ approach a shelf region where the maximum shelf wave frequency $\sigma_M$ for their particular mode is near $\sigma$ [Smith, 1975]. If shelf variations cause $\sigma_M$ to fall well below $\sigma$, then presumably the waves are totally reflected, with large amplitudes near where $\sigma_M = \sigma$.

Poleward propagation (varying f) in stratified conditions has been studied by Allen and Romea [1980] and numerically by Suginohara [1981]. Near the equator, f is small, $N^2H^2/f^2L^2$ (effective stratification) is large, and internal Kelvin-like waves are expected (section 5). As f increases poleward, the waves evolve continuously to less stratified, more topographic shelf wave forms, provided the evolution distance is long compared with the wave's longshore scale. (Two-layer coastal-wall models may suggest spuriously rapid transition near the latitude where "pure" internal Kelvin wave and shelf wave speeds are equal, and hence spurious difficulty for the traveling wave in effecting the transition.)

The Coriolis parameter is special in that slow longshore variations support other (Rossby) waves offshore. Internal Kelvin and shelf waves subject to varying f may slowly lose energy to the ocean interior as Rossby waves [McCreary, 1976; Suginohara, 1981], particularly at low frequencies $\sigma$ when the Rossby wave scale (earth's radius times $\sigma/f$) matches the shelf wave scale (shelf width times $f/\sigma$).

## Small Features

Kelvin wave response to small (respectively $\epsilon(gH)^{1/2}/f$, $\epsilon H$) irregularities of finite extent, in an otherwise straight coast or uniform depth H, were considered by Pinsent [1972] and Miles [1972, 1973]. Effects are generally $O(\epsilon^2)$, but $O(\epsilon)$ near the irregularity, and phase shifts after depth changes are $O(\epsilon)$. Chao [1980] found that a small bottom bump in continuous stratification scatters some energy from a Kelvin wave into all internal Kelvin waves, especially the first (n = 1 in (22)); disturbances local to the bump form in $\sigma < |f|$, while energy is scattered away from the coast in $\sigma > |f|$. Killworth [1978] considered the effects on internal Kelvin waves of coastal curvature (relatively unimportant) and longshore depth variations. Scattering into modes with nearby n in (22) is preferred. A mode-n pulse (as opposed to a wave train) leaves steady flow over a canyon or spur, as well as scattering into other modes, especially n + 1.

Barotropic shelf waves on an exponential depth profile have been considered in the presence of coastal bumps [Buchwald, 1977] and bottom bumps [Chao et al., 1979]. Energy is scattered forward and backward into all possible modes at the frequency $\sigma$ of the incident wave. The scattered waves are $\pm 90°$ out of phase. The highest possible scattered mode (m, say) is particularly energetic; if $\sigma$ is close to its maximum frequency so that $\partial\sigma_m/\partial k \sim 0$ at $\sigma_m(k) = \sigma$, then its energy disperses very slowly and there is a resonanat response. However, it should be noted that smoother topographic features, regarded as compositions of bumps, result in a superposition of scattered waves; scattering to lower wave numbers (i.e., low modes and forward scattering of energy) is favored as higher wave numbers tend to self-cancel.

For long ($L_W$) shelf waves and comparably long ($L_T$) bumps on an exponential depth profile, Allen [1976b] found that the waves adopt the appropriate local form. They also scatter, especially higher modes; rates are generally of the order of $L_W/L_T$ if $L_W/L_T \ll 1$ or $(L_T/L_W)^2$ if $L_T/L_W \ll 1$.

## Shelf Waves Over Larger Features

If the depth profile has the self-similar form $h((x-c(y))/L(y))$, then Hsueh [1980] has shown that long (much longer than the shelf width L(y)) shelf waves propagate with changes of amplitude but no scattering or change of form, provided c and L also vary slowly with y (c/c', L/L' $\gg$ L). Stronger currents and shorter wavelengths are implied on narrow sections of shelf.

Numerical calculations of Wang [1980] show stronger currents where depth contours have converged, for shorter scales alongshore. At low frequencies, this is to be expected, as the flow is guided along depth contours (cf. section 6), with little reflection or scattering. The converged contours in the calculations supported a lower maximum shelf wave frequency $\sigma_M$; as the incident wave frequency $\sigma$ approached $\sigma_M$, there was increasing backscatter of energy, with total reflection for $\sigma > \sigma_M$. For waves passing over a canyon or spur, there was forwardscatter and backscatter of energy, as suggested by the small-bump models.

## Small Random Variations

Kelvin waves suffer reduced speed and energy flux along an irregular (but basically straight) coastline [Mysak and Tang, 1974]. For $\sigma > |f|$, there is a slow loss of energy to the ocean interior, as also occurs for edge waves [Fuller and Mysak, 1977]. Then the wave crests tilt toward the coast.

Shelf waves on a randomly perturbed exponential depth profile in a channel were considered by Brink [1980]. Scattering into other shelf waves is particularly strong for frequencies corresponding to $\partial\sigma/\partial k = 0$ for some mode and causes attenuation and a change of speed of the original wave.

Random-variation analyses use ensemble averages, and application involves the ergodic hypothesis. Hence attenuation (say) is not necessarily implied in any particular context. The results may be applied with most confidence to overall effects of long shelves with stationary statistics.

## Abrupt Features

Most analyses have been for Kelvin waves in water of uniform depth. For $\sigma < |f|$, all energy remains in the wave, but phase shifts are possible; for $\sigma > |f|$, some energy is generally lost to the ocean interior. LeBlond and Mysak [1978, pp. 266-268] review models of diffraction around a thin semi-infinite barrier, through a narrow gap in an infinite barrier (gap width $w \ll (gh)^{1/2}/\sigma$ but the amplitude diffracted through the gap is relatively large, $O[2/\ln(gh/\sigma^2 w^2)]$), and around corners. Interior corners (angle less than $\pi$) advance the Kelvin wave phase, and special angles $\pi/(2n + 1)$ give perfect energy transmission even in $\sigma > |f|$, as do successive corners forming the head of all but the widest gulfs [e.g., Pearson, 1977]. Green's functions for generating Kelvin waves by coastal and offshore forcing [Buchwald, 1971; Keller and Watson, 1981] are useful for some geometries (e.g., the narrow gap). We remark that these solutions are relevant specifically for the shelf (depth $h$, width L) at high enough frequencies $\sigma > (gh)^{1/2}/L$ so that the Kelvin wave is on the shelf (section 3). Several models treat a shelf of finite extent [e.g., Huthnance, 1980]; successive reflections of progressive trapped modes may synthesize near-resonant standing waves.

Shelf wave transmission, and scattering into other modes, at a narrow peninsula extending partly across an exponential-profile shelf, has been treated by Davis [1981]. A complete barrier across the shelf implies reflection into (short, slow) waves of opposite group velocity synthesizing minus the incident longshelf velocity.

## 9. Generation and Dissipation

Forcing of shelf seas may be via any of the boundaries: the seafloor, ocean, sea surface, or coastline. Shelf waves may also be generated from dynamical imbalances or perturbations of other "preexisting" motion.

Seafloor movements, or earthquakes, are important but of relatively short periods (minutes) and are outside the scope of this discussion.

Oceanic motions may impinge on the continental slope. Exponential slopes have been studied by Kroll and Niiler [1976], by Kroll [1979] (in relation to wind forcing) and numerically by Ou and Beardsley [1980] including stratification. Penetration onto the shelf, usually limited by reflections at breaks in the depth profile, occurs for particular slope widths (relative to the wavelength); then friction generally prevents penetration to the coast [Kroll and Niiler, 1976]. At the equator, waves propagating eastward across the ocean and reaching the coast generate northward and southward propagating Kelvin and internal Kelvin waves [Moore and Philander, 1977]. All shelf-sea tidal models, too numerous to mention, in effect study the shelf-sea response to ocean tides $O(0.1 - 1 \text{ m})$; Kelvin waves, at least, play a major role. Ocean-shelf interaction has come to be recognized as of the greatest significance, and is discussed by Smith [1983].

Atmospheric pressure forcing the sea surface can be quite effective at generating Kelvin and edge waves with a strong surface gravity wave element. Resonance is possible in principle if the pressure field matches the speed and wave number (!) of a wave mode along the shelf. More realistically, a front traveling faster than the slowest (lowest) mode generates an edge wave wake with amplitudes a few times larger than the atmospheric pressure change. To be slower than the front, the gravity wave speed $O(gh)^{1/2}$ must be for shelf-sea (not oceanic) depths h; hence the frequency must be quite high: $\sigma > (gh)^{1/2}/L$ (section 3). A wedge [Greenspan, 1956], step shelf, and exponential depth profiles have been studied [see Viera and Buchwald, 1982].

The atmospheric pressure generation of shelf waves was considered by Robinson [1964] and Mysak [1967] in connection with early observations correlating pressure and sea level.

Longshore wind stress is now believed to be most effective for generating shelf waves adjacent to a coast, as first expounded by Adams and Buchwald [1969] for an exponential depth profile and low frequencies. Thomson [1970] and Kajiura [1962] considered the generation of Kelvin and internal Kelvin waves, the latter contributing larger currents in the ratio of lower to upper layer depths in a two-layer model. At low latitudes, Romea and Allen [1982] find that wind stress excites the internal Kelvin wave of a two-layer model, rather than shelf waves; at mid-latitudes the shelf circulation is generally responsive to wind stress through shelf waves. A first-order wave equation governs the developing amplitude of a nondispersive ($\sigma \ll |f|$) Kelvin or shelf wave as it progresses along the coast forced by the local wind at each stage; Gill and Schumann [1974] and Gill and Clarke [1974] treated unstratified and stratified contexts. When or where the forcing ceases, the wave propagates onward and is then most recognizable.

Allen [1980] reviews wind-driven currents on the shelf. The simplest dynamical approach, from (1) integrated throughout the water depth, is that the surface longshore wind stress $\underset{\sim}{\tau}$ accelerates the longshore transport hu. A more sophisticated view, integrating (1) through the upper layer where $\underset{\sim}{\tau}$ acts, is that at low frequencies, $\underset{\sim}{\tau}$ induces a cross-shelf surface transport $\underset{\sim}{\tau}/\rho f \times \underset{\sim}{k}$; coastal blocking induces a compensating return flow beneath, which is acted upon by the Coriolis force to give the same accelerating longshore transport. A typical stress of $0.1 \text{ N m}^{-2}$ acting for $10^5$ s (~ 1 day) accelerates a 100 m depth of water to $0.1 \text{ m s}^{-1}$.

Winds blowing across the depth contours may be more effective if the coast is very distant, as found by Brink [1983] for ridge topography. Cross-scarp winds forcing double Kelvin waves were considered by Mysak [1969].

Longshore shelf variations encourage wave generation. Scattering from waves already present has been treated in section 8; frequency maxima of the various modes appear to be favored. For $\sigma > |f|$, waves incident from the ocean interior generate Kelvin waves along an irregular coastline [Pinsent, 1972; Howe and Mysak, 1973]; in $\sigma < |f|$, shelf waves can be generated from (say) oceanic tides [Huthnance, 1974]. Clarke [1977a] emphasized that a curving coast varies the longshore wind stress component, encouraging a more wavelike response. Martell and Allen [1979] found that a wind-accelerated longshore current over small topographic variations develops lee waves, which are carried downstream when the current exceeds the maximum (opposing) phase speed.

Other mechanisms generating shelf waves may include large and varying river runoff [Wolanski and van Senden, 1983] or a co-oscillating gulf; each provides a coastal source of water similar in effect to the onshore transport $\underline{\tau}/\rho f \times \underline{k}$ associated with a localized longshore wind stress $\underline{\tau}$. Thomson and Crawford [1982] describe a mechanism for weakly generating a first-mode shelf wave from bottom-frictional effects on a Kelvin wave.

Friction causes cross-shelf phase shifts and significant damping of coastal trapped waves. Integrating (1) through depth, with bottom drag $\rho r \underline{u}$ and neglecting $\int u \, dz$ (by continuity) and $\partial p/\partial y$,

$$\partial v/\partial t + rv/h = \tau/\rho h$$

suggesting that v lags $\tau$ less for smaller frequencies and depths and greater friction. Accordingly, Brink and Allen [1978] found that for any one forced wave mode, nearshore currents lag the wind less than do currents in deeper water offshore. The total response may be more complex, particularly if one constituent forced mode is near resonance [Simons, 1983; Brink and Allen, 1983]. Frictional generation of other wave modes also induces cross-shelf phase shifts in a stratified context [Brink 1982a]. Damping rates may be estimated as $r/h = O(3 \times 10^{-3} \times U/h)$, where U is the largest (e.g., tidal) current present (see, for example, Heaps [1978]). Hence the decay time is less than 4 days for $U = 0.1$ m s$^{-1}$, $h = 100$ m, corresponding to a decay distance of 300 km for a wave propagating at 1 m s$^{-1}$. The decay distance is generally less for the wave energy, propagating at the group velocity. Evidently, frictional effects are most important for long-period and slowly propagating (higher mode or short wavelength) waves. Friction is locally dominant at the coast, where $h \to 0$. However, wave modes decay only at rates $O(r/H)$, where H is a typical depth [Hukuda and Mysak, 1982]. Much less decay is predicted for waves over the deeper continental slope, particularly if (dubiously!) only interior friction is included. Martinsen and Weber [1981] demonstrate the contrast for internal Kelvin waves.

Losses to Rossby waves are possible when f varies alongshore (section 8).

10. Observations

Shelf waves have been widely observed. Adding the Weddell Sea observations of Middleton et al. [1982] to the list in the work by Mysak [1980], coastlines of various orientations and all continents in both the northern and southern hemispheres are included.

The lowest mode, calculated for unstratified conditions, has been most often identified. Its peak coastal elevation is relatively easily and frequently measured. However, additional offshore current measurements are valuable for detecting the presence and form of higher modes [Hsieh, 1982a]. Currents off Oregon are thought to show modes 1, 2, and 3 at various times [Cutchin and Smith, 1973; Hsieh and Mysak, 1980; Hsieh, 1982b]. Current measurements from the Middle Atlantic Bight were represented empirically by Ou et al. [1981] as two plane waves, separately identified as wind forced (then adjusted to match the wind field translation) and free. The free wave speed agreed reasonably with independent estimates from coastal elevations [Wang, 1979].

Stratification may significantly affect the waves' structure, as indicated by seasonal changes in the vertical structure and offshore scale of currents on the Oregon shelf [Huyer et al., 1978]. In accord with calculations, stratification reduces the offshore decay of upper level currents in the lowest shelf wave mode observed off Vancouver Island [Crawford and Thomson, 1982]. The clearest observations of internal Kelvin waves appear to have been in the Great Lakes [see Clarke, 1977b] and around Bermuda [Hogg, 1980], where the bottom slopes steeply with no real shelf. Equivalently, the offshore decay scale $c_n/f$ (section 5) is large near the equator; current, temperature, and coastal sea level records off Peru between 10°S and 15°S [Smith, 1978] showed fluctuations with an offshore scale of 70 km, greater than the shelf width. The fluctuations were not well correlated with local winds but propagated poleward at 200 km d$^{-1}$ from north of 10°S, resembling free internal Kelvin waves (with some modification by the shelf). Identification of bottom-trapped waves requires knowledge of the local slope and stratification and may be difficult despite the apparent prevalence of near-bottom currents. Thompson and Luyten [1976] and Hogg [1981] present evidence from the eastern U.S. continental slope. There is some evidence of enhanced energy at the maximum bottom-trapped wave frequency $Nh' > f$ on the NW Africa continental slope [e.g., Huthnance and Baines, 1982].

Tides include Kelvin waves around many of the world's coasts. Semidiurnal models fitting observations for California [Munk et al., 1970], Scotland [Cartwright et al., 1980], and NW Africa [Huthnance and Baines, 1982] have a dominant Kelvin wave. However, measurements to the west of Scotland [Cartwright et al., 1980] and Vancouver Island [Crawford and Thomson, 1982] clearly reveal a dominant shelf wave contribution to diurnal tidal currents. (As a historical note, diurnal tidal currents between the Scottish islands were the first recorded shelf wave [Moray, 1665]. Although not then understood, the currents were remarked because semidiurnal tidal elevations predominate by factors of 10 to 20 locally).

Shelf waves with frequencies $\sigma$ near the maximum $\sigma_M$ for some mode (i.e., near-zero group velocity $\partial\sigma/\partial k$) appear in several observations. The Scottish diurnal tide shelf wave may be strong because the shelf profile changes northward, $\sigma_M$ decreasing to near $\sigma$ [Smith, 1975]. Brooks [1978] found increased coherence, at frequencies near $\sigma_M$, between sea levels at two North Carolina ports about 125 km apart and between the sea levels and atmospheric variables (pressure and two wind stress components).

Mean currents figure in many observations, motivating the analyses in section 6. Adjacent to the Florida current, temperature and coastal sea-level fluctuations have taken the form of wind-forced shelf waves [Brooks and Mooers, 1977b]. Gulf Stream meanders have been interpreted as barotropically unstable shelf waves advected from the Blake Plateau [Niiler and Mysak, 1971]. Shelf currents in Shelikof Strait, Alaska, showed peak energies at frequencies which were baroclinically unstable according to a two-layer model [Mysak et al., 1981b]; with due regard to seasonal changes of stratification, calculated wavelengths at the observed frequencies roughly matched observed wavelengths. Such conclusions depend on rather simple models (section 6); although consistency was found in these cases, modeling uncertainties prevent confidence in general. For example, the Norwegian current is observed to be unstable, with a wavelength consistent with a baroclinic instability model, but the overall energy budget for the fluctuations (including barotropic stability/instability) suggests stability [Schott and Bock, 1980].

## 11. The Role of Shelf Waves

Shelf-sea motions are frequently dominated by tides and/or currents driven by the wind stress over the shelf. The wavelike nature of the tides is clear, and models in terms of Kelvin waves date from Taylor's [1920] use of a rectangular gulf to represent the North Sea. Recent models [Cartwright et al., 1980; Crawford and Thomson, 1982] demonstrate the dominance of the lowest-mode shelf wave in diurnal tidal currents at different locations. Storm surges may also appear wavelike, but it has only recently been appreciated that waves also play a role in steadier wind-driven currents.

Longshore winds accelerate longshore flow (section 9), over the whole shelf width for typical extensive wind fields. On most shelves, with relatively weak stratification (section 5), the induced flow represents forcing principally of the lowest shelf wave mode (having one-signed velocity over the shelf width). Within the forcing region the motion tends to match the wind field and may not appear wavelike [Hickey, 1981] or may apparently propagate faster than the free wave [Chao, 1981], but the wave's presence implies propagation away from the forcing region. The wind's influence is therefore felt "downstream" (in the sense of wave propagation); moreover, within the forcing region the response tends to be greater "downstream."

In stratified conditions the wind may induce upwelling if surface waters are blown offshore so that cooler waters are drawn up from below. Such motion also corresponds to a (stratified) wave mode or combination of modes. "Downstream" influence again follows. Suginohara's [1982] numerical calculations provide a clear demonstration.

The wave's "viewpoint" is that, as it propagates, its amplitude is continually incremented according to the local forcing [Gill and Schumann, 1974; Gill and Clarke, 1974].

For any fixed position on the shelf, the motion results from local forcing together with motion propagating from forcing regions "upstream." Major implications are (1) that great care is needed in using models assuming zero longshore gradients and (2) that the influence of "upstream" boundary conditions may penetrate far into a model [e.g., Beardsley and Haidvogel, 1981].

The formation and maintenance of steady currents are similarly influenced. Friction introduces a "downstream" decay length for shelf waves. Considering ever lower frequencies, shelf waves retain their form, speed and corresponding decay length in a continuous transition to solutions of a parabolic equation

$$r(h^{-2}\psi')' + (f/h)' \, \partial\psi/\partial y = 0 \qquad (24)$$

replacing (15) (for $\partial\tau/\partial z = -r\underset{\sim}{u}/h$ in (1) and relatively long scales alongshore). Csanady [1978] describes solutions in the case of uniform slope. Steady currents decay "downstream," the component shelf wave patterns each decaying in the corresponding shelf wave decay length. We still look "upstream" for forcing, but only as far as the decay length; currents cannot be "pushed" from greater distances. This restriction perhaps applies more to currents on the shelf than along the slope, where reduced friction effects (due to greater depth h, at least, (cf. section 9)) imply a greater decay length.

Oceanic motions have to accommodate to the presence of the coast and shelf. In the simplest case of uniform depth and a coastal wall, internal Kelvin waves are generated as the various vertical structure modes (e.g., $\cos n\pi z/h$; see (22)) of the normal velocity come to zero at the wall. For other shelf profiles and low frequencies (at least), stratified shelf wave forms perform this role of adjusting oceanic currents to the details of the shelf profile [Allen, 1976a; Huthnance

1978b]. All the above implications of decaying "downstream" propagation and influence carry over to this case. For many shelves, with relatively weak stratification and a majority of modes active over the shelf and upper slope only, the minority of modes active over the remaining slope may gain prominence in this role; a barotropic calculation by Wang [1982] shows concentrated flow along the continental slope (not the shelf).

There is an analogy with McCreary's [1981] model for adjustment of wind-driven flow, in a stratified constant-depth ocean, to the eastern boundary. However, in his case, confinement of the vertical structure modes against (only) the eastern boundary (resulting in a poleward under-current) is due to density diffusion.

The response to, for example, oceanic longshore pressure gradients, oceanic eddies, and Gulf Stream meanders impinging on the continental slope is also of interest [Smith, 1983].

Natural modes of the ocean are significantly affected by the shelves. In general terms, modes depending on $f/h$ gradients have increased frequencies and concentrate over the shelves (and other topography) [Ripa, 1978]. Numerical calculations by Platzman et al. [1981] show 13 modes with periods between 30 and 80 hours, each mode being localized over one shelf area.

## 12. Outlook

There remains scope for further analytical and numerical studies of shelf waves. Perhaps the largest area of ignorance concerns stratified shelf seas above the inertial frequency $|f|$. Chapman [1982a], neglecting rotation, finds nearly trapped analogues of edge waves over a step shelf and uniform stratification; we wish to know whether there are such analogues in other contexts, whether short-wavelength forms are bottom trapped (as below the inertial frequency), and the character of those modes whose frequency $\sigma$ increases through $|f|$ as their wave number $k$ increases. The many studies of internal tides with zero longshelf wavenumber $k$ [Baines, this volume] treat appropriate frequencies but do not describe dispersion and energy propagation along the shelf.

Our description of the role of shelf waves is supported at present only by analysis for $\sigma \ll |f|$. Owing to the importance of friction, wave propagation is likely to be more apparent at higher frequencies, $\sigma = O(|f|)$, which warrant more attention, being associated also with some of the strongest wind forcing, e.g., storms.

More systematic calculations are needed in the context of continuous stratification and mean current shears (horizontal and vertical), relating the stable and unstable wave modes to the potential vorticity distribution and to the modes when mean currents are zero. Are there as many modes as two-layer models tend to suggest? There is also scope for more calculations of waves in realistic contexts, for example, on the shelf/slope water front bordering many shelves.

Ideas about the expected distribution between wave modes, frequencies, and wavelengths have not been synthesized. Low modes may be preferred by direct forcing, but nonlinearities and irregular shelves scatter to higher modes, especially at modal frequency maxima. Perhaps strong friction prevents the growth of significant scattered wave energy. Losses to Rossby waves in the ocean interior have not been evaluated.

Observations of shelf waves, although numerous, are not all of good quality. Particularly striking is the lack of laboratory experiments for stratified shelf seas or to test any of the theoretical ideas regarding nonlinear effects, scattering into other modes by shelf variations, or shelf wave generation. Unfortunately, field experiments to test modal forms and dispersion should preferably include several current records in a cross-shelf section, and measurements at several points along the shelf, over a period of some months. The required effort increases further if scattered modes or oceanic forcing are to be distinguished. A recurring problem is the sparsity of good wind data over the sea, required to separate forced and free waves [Hickey, 1981]. Nevertheless, recent experiments satisfying these criteria have been undertaken off Nova Scotia, Vancouver Island, California, Scotland, and New South Wales, and clearer results are beginning to emerge.

On real shelves, circulation may only be well simulated by numerical models (the subject of a future volume in this series). Such models must compromise coverage for resolution and generally require prescribed conditions at some open (e.g., shelf-edge and cross-shelf) boundaries and an initial time. The choice of model boundaries and conditions may be guided (section 11) by our understanding of shelf waves as natural modes.

## References

Adams, J. K., and V. T. Buchwald, The generation of continental shelf waves, J. Fluid Mech., 35, 815-826, 1969.

Allen, J. S., Coastal trapped waves in a stratified ocean, J. Phys. Oceanogr., 5, 300-325, 1975.

Allen, J. S., On forced long continental shelf waves on an f-plane, J. Phys. Oceanogr., 6, 426-431, 1976a.

Allen, J. S., Continental shelf waves and along-shore variations in bottom topography and coastline, J. Phys. Oceanogr., 6, 864-878, 1976b.

Allen, J. S., Models of wind-driven currents on the continental shelf, Annu. Rev. Fluid Mech., 12, 389-433, 1980.

Allen, J. S., and R. D. Romea, On coastal trapped waves at low latitudes in a stratified ocean, J. Fluid Mech., 98, 555-585, 1980.

Baines, P. G., Internal tides, internal waves, and near-inertial motions, this volume.

Bane, J. M., Coastal-trapped and frontal-trapped

waves in a baroclinic western boundary current, J. Phys. Oceanogr., 10, 1652-1668, 1980.

Barton, N. G., Resonant interactions of shelf waves with wind-generated effects, Geophys. Astrophys. Fluid Dyn., 9, 101-114, 1977.

Barton, N. G., and V. T. Buchwald, The non-linear generation of shelf waves, in Waves on Water of Variable Depth, Lecture Notes Phys., vol. 64, edited by D. G. Provis and R. Radok, pp. 194-201, Springer-Verlag, New York, 1977.

Beardsley, R. C., and D. B. Haidvogel, Model studies of the wind-driven transient circulation in the Middle Atlantic Bight, 1, Adiabatic boundary conditions, J. Phys. Oceanogr., 11, 355-375, 1981.

Bennett, J. R., A theory of large-amplitude Kelvin waves, J. Phys. Oceanogr., 3, 57-60, 1973.

Bondok, S. A. el W., Some effects of varying depth on trapped waves and ocean currents in one and two-layer oceans, Ph.D. thesis, 310 pp., Univ. of London, 1980.

Brink, K. H., Propagation of barotropic continental shelf waves over irregular bottom topography, J. Phys. Oceanogr., 10, 765-778, 1980.

Brink, K. H., The effect of bottom friction on low-frequency coastal trapped waves, J. Phys. Oceanogr., 12, 127-133, 1982a.

Brink, K. H., A comparison of long coastal trapped wave theory with observations off Peru, J. Phys. Oceanogr., 12, 897-913, 1982b.

Brink, K. H., Low frequency free wave and wind-driven motions over a submarine bank, J. Phys. Oceanogr., 13, 103-116, 1983.

Brink, K. H., and J. S. Allen, On the effect of bottom friction on barotropic motion over the continental shelf, J. Phys. Oceanogr., 8, 919-922, 1978.

Brink, K. H., and J. S. Allen, Reply, J. Phys. Oceanogr., 13, 149-150, 1983.

Brooks, D. A., Subtidal sea level fluctuations and their relation to atmospheric forcing along the Carolina coast, J. Phys. Oceanogr., 8, 481-493, 1978.

Brooks, D. A., and C. N. K. Mooers, Free, stable continental shelf waves in a sheared, barotropic boundary current, J. Phys. Oceanogr., 7, 380-388, 1977a.

Brooks, D. A., and C. N. K. Mooers, Wind-forced continental shelf waves in the Florida current, J. Geophys. Res., 82, 2569-2576, 1977b.

Buchwald, V. T., Long waves on oceanic ridges, Proc. R. Soc. London, Ser. A, 308, 343-354, 1969.

Buchwald, V. T., The diffraction of tides by a narrow channel, J. Fluid Mech., 46, 501-511, 1971.

Buchwald, V. T., Diffraction of shelf waves by an irregular coastline, in Waves on Water of Variable Depth, Lecture Notes Phys., vol. 64, edited by D. G. Provis and R. Radok, pp. 188-193, Springer-Verlag, New York, 1977.

Caldwell, D. R., D. L. Cutchin, and M. S. Longuet-Higgins, Some model experiments on continental shelf waves, J. Mar. Res., 30, 39-55, 1972.

Cartwright, D. E., J. M. Huthnance, R. Spencer, and J. M. Vassie, On the St. Kilda shelf tidal regime, Deep Sea Res., 27, 61-70, 1980.

Chao, S.-Y., Topographic coupling of surface and internal Kelvin waves, J. Phys. Oceanogr., 10, 1147-1158, 1980.

Chao, S.-Y., Forced shelf circulation by an along-shore wind band, J. Phys. Oceanogr., 11, 1325-1333, 1981.

Chao, S.-Y., L. J. Pietrafesa and G. S. Janowitz, The scattering of continental shelf waves by an isolated topographic irregularity, J. Phys. Oceanogr., 9, 687-695, 1979.

Chapman, D. C., Nearly trapped internal edge waves in a geophysical ocean, Deep Sea Res., 29A, 525-533, 1982a.

Chapman, D. C., On the failure of Laplace's tidal equations to model subinertial motions at a discontinuity in depth, Dyn. Atmos. Oceans, 7, 1-16, 1982b.

Chapman, D. C., On the influence of stratification and continental shelf and slope topography on the dispersion of sub-inertial coastally-trapped waves. J. Phys. Oceanogr., 13, 1641-1652, 1983.

Chapman, D. C., and M. C. Hendershott, Shelf wave dispersion in a geophysical ocean, Dyn. Atmos. Oceans, 7, 17-31, 1982.

Clarke, A. J., Wind-forced linear and nonlinear Kelvin wavs along an irregular coastline, J. Fluid Mech., 83, 337-348, 1977a.

Clarke, A. J., Observational and numerical evidence for wind-forced coastal trapped long waves, J. Phys. Oceanogr., 7, 231-247, 1977b.

Clarke, D. J., Long edge waves over a continental shelf, Dtsch. Hydrogr. Z., 27, 1-8, 1974.

Collings, I. L., and R. Grimshaw, The effect of topography on the stability of a barotropic coastal current, Dyn. Atmos. Oceans, 5, 83-106, 1980.

Crawford, W. R., and R. E. Thomson, Continental shelf waves of diurnal period along Vancouver Island, J. Geophys. Res., 87, 9516-9522, 1982.

Csanady, G. T., The arrested topographic wave, J. Phys. Oceanogr., 8, 47-62, 1978.

Cutchin, D. L., and R. L. Smith, Continental shelf waves: Low-frequency variatons in sea level and currents over the Oregon continental shelf, J. Phys. Oceanogr., 3, 73-82, 1973.

Davis, A. M. J., The scattering by a peninsula of the dominant continental shelf wave, Philos. Trans. R. Soc. London, Ser. A, 303, 383-431, 1981.

Denbo, D. W., and J. S. Allen, Mean flow generation on a continental margin by periodic wind forcing, J. Phys. Oceanogr., 13, 78-92, 1983.

deSzoeke, R. A., Some effects of bottom topography on baroclinic stability, J. Mar. Res., 33, 93-122, 1976.

Fuller, J. D., and L. A. Mysak, Edge waves in the presence of an irregular coastline, J. Phys. Oceanogr., 7, 846-855, 1977.

Gill, A. E., and A. J. Clarke, Wind-induced upwelling, coastal currents and sea-level changes, Deep Sea Res., 21, 325-345, 1974.

Gill, A. E., and E. H. Schumann, The generation of long shelf waves by the wind, J. Phys. Oceanogr., 4, 83-90, 1974.

Green, K., and V. T. Buchwald, Interior shelf waves on an equatorial β-plane, J. Phys. Oceanogr., 9, 1299-1304, 1979.

Greenspan, H. P., The generation of edge waves by a moving pressure disturbance, J. Fluid Mech., 1, 574-592, 1956.

Griffiths, R. W., and P.F. Linden, The stability of buoyancy-driven coastal currents, Dyn. Atmos. Oceans, 5, 281-306, 1981.

Griffiths, R. W., and P. F. Linden, Laboratory experiments on fronts, I, Density-driven boundary currents, Geophys. Astrophys. Fluid Dyn., 19, 159-187, 1982.

Grimshaw, R. Nonlinear aspects of long shelf waves, Geophys. Astrophys. Fluid Dyn., 8, 3-16, 1977a.

Grimshaw, R. The stability of continental shelf waves, I, Side band instability and long wave resonance, J. Aus. Math. Soc., B20, 13-30, 1977b.

Grimshaw, R. The effects of a variable Coriolis parameter, coastline curvature and variable bottom topography on continental shelf waves, J. Phys. Oceanogr., 7, 547-554, 1977c.

Grimshaw, R. The effect of a mean current on Kelvin waves, J. Phys. Oceanogr., 13, 43-53, 1983.

Hall, R. E., A note on a semicircle theorem, Dyn. Atmos. Oceans, 5, 113-121, 1980.

Heaps, N. S., Linearized vertically-integrated equations of residual circulation in coastal seas, Dtsch. Hydrogr. Z., 31, 147-169, 1978.

Hickey, B. M., Alongshore coherence on the Pacific Northwest continental shelf (January-April, 1975), J. Phys. Oceanogr., 11, 822-835, 1981.

Hogg, N. G., Observations of internal Kelvin waves trapped round Bermuda, J. Phys. Oceanogr., 10, 1353-1376, 1980.

Hogg, N. G., Topographic waves along 70°W on the continental rise, J. Mar. Res., 39, 627-649, 1981.

Howe, M. S., and L. A. Mysak, Scattering of Poincaré waves by an irregular coastline, J. Fluid Mech., 57, 111-128, 1973.

Hsieh, W. W., On the detection of continental shelf waves, J. Phys. Oceanogr., 12, 414-427, 1982a.

Hsieh, W. W., Observations of continental shelf waves off Oregon and Washington, J. Phys. Oceanogr., 12, 887-896, 1982b.

Hsieh, W. W., and L. A. Mysak, Resonant interactions between shelf waves, with applications to the Oregon shelf, J. Phys. Oceanogr., 10, 1729-1741, 1980.

Hsueh, Y., Scattering of continental shelf waves by longshore variations in bottom topography, J. Geophys. Res., 85, 1147-1150, 1980.

Hukuda, H., Subharmonic destabilisation off Vancouver Island, J. Phys. Oceanogr., 12, 285-292, 1982.

Hukuda, H., and L. A. Mysak, On the damping of second-class waves on a sloping beach, J. Phys. Oceanogr., 12, 1527-1531, 1982.

Huthnance, J. M., On the diurnal tidal currents over Rockall Bank, Deep Sea Res., 21, 23-35, 1974.

Huthnance, J. M., On trapped waves over a continental shelf, J. Fluid Mech., 69, 689-704, 1975.

Huthnance, J. M., On coastal trapped waves: Analysis and numerical calculation by inverse iteration, J. Phys. Oceanogr., 8, 74-92, 1978a.

Huthnance, J. M., On coastal trapped wave response to wind over the deep ocean, Ocean Modelling, 17, 1-3, 1978b.

Huthnance, J. M., On shelf-sea "resonance" with application to Brazilian M3 tides, Deep Sea Res., 27, 347-366, 1980.

Huthnance, J. M., On mass transports generated by tides and long waves, J. Fluid Mech., 102, 371-391, 1981.

Huthnance, J. M., and P. G. Baines, Tidal currents in the northwest African upwelling region, Deep Sea Res., 29A, 285-306, 1982.

Huyer, A., R. L. Smith, and E. J. C. Sobey, Seasonal differences in low-frequency current fluctuations over the Oregon continental shelf, J. Geophys. Res., 83, 5077-5089, 1978.

Iida, H., Edge waves on the linearly sloping coast, I, Free waves, Oceanogr. Mag., 22, 37-62, 1970.

Kajiura, K., A note on the generation of boundary waves of the Kelvin type, J. Oceanogr. Soc. Jpn., 18, 51-58, 1962.

Kawabe, M., Coastal trapped waves in a two-layer ocean, J. Oceanogr. Soc. Jpn., 38, 115-124, 1982.

Keller, J. B., and J. G. Watson, Kelvin wave production, J. Phys. Oceanogr., 11, 284-285, 1981.

Kenyon, K. E., Edge waves with current shear, J. Geophys. Res., 77, 6599-6603, 1972.

Killworth, P. D., Coastal upwelling and Kelvin waves with small longshore topography, J. Phys. Oceanogr., 8, 188-205, 1978.

Kimura, R., Barotropic instability of a boundary jet on a sloping bottom, Geophys. Fluid Dyn., 7, 205-230, 1976.

Kroll, J., The kinetic energy on a continental shelf from topographic Rossby waves generated off the shelf, J. Phys. Oceanogr., 9, 712-723, 1979.

Kroll, J., and P. P. Niiler, The transmission and decay of barotropic topographic Rossby waves incident on a continental shelf, J. Phys. Oceanogr., 6, 432-450, 1976.

Lacombe, H., Trapping of waves by a constant slope internal interface in a two-layer ocean, J. Phys. Oceanogr., 12, 337-360, 1982.

Lamb, H., Hydrodynamics, 738 pp., Cambridge University Press, New York, 1932.

LeBlond, P. H., and L. A. Mysak, Waves in the Ocean, 602 pp., Elsevier, New York, 1978.

Longuet-Higgins, M. S., On the trapping of long-period waves round islands, J. Fluid Mech., 37, 773-784, 1969.

Longuet-Higgins, M. S., Topographic Rossby waves, Mem. Soc. R. Sci. Liege, Ser. 6, 2, 11-16, 1972.

Louis, J. P., Low frequency edge waves over a trench-ridge topography adjoining a straight coastline, Geophys. Astrophys. Fluid Dyn., 9, 229-239, 1978.

Lozano, C., and R. E. Meyer, Leakage and response of waves trapped round islands, Phys. Fluids, 19, 1075-1088, 1976.

Luther, M. E., and J. M. Bane, Coastal-trapped waves in a continuously stratified western boundary current, Ocean Modelling, 25, 6-8, 1979.

Martell, C. M., and J. S. Allen, The generation of continental shelf waves by alongshore variations in bottom topography, J. Phys. Oceanogr., 9, 696-711, 1979.

Martinsen, E. A., and J. E. Weber, Frictional influence on internal Kelvin waves, Tellus, 33, 402-410, 1981.

McCreary, J. P., Eastern tropical ocean response to changing wind systems--With appplication to El Niño, J. Phys. Oceanogr., 6, 632-645, 1976.

McCreary, J. P., A linear stratified ocean model of the coastal undercurrent, Philos. Trans. R. Soc., London, Ser. A, 302, 385-413, 1981.

McKee, W. D., Critical levels for forced Rossby and continental shelf waves, Pure Appl. Geophys., 117, 841-850, 1979.

Mechoso, C. R., Baroclinic instability of flows along sloping boundaries, J. Atmos. Sci., 37, 1393-1399, 1980.

Mechoso, C. R., and D. M. Sinton, Instability of baroclinic flows with horizontal shear along topography, J. Phys. Oceanogr., 11, 813-821, 1981.

Middleton, J. H., T. D. Foster, and A. Foldvik, Low frequency currents and continental shelf waves in the southern Weddell Sea, J. Phys. Oceanogr., 12, 618-634, 1982.

Miles, J. W., Kelvin waves on oceanic boundaries, J. Fluid Mech., 55, 113-127, 1972.

Miles, J. W., Kelvin-wave diffraction by changes in depth, J. Fluid Mech., 57, 401-413, 1973.

Miles, J. W., and F. K. Ball, On free-surface oscillations in a rotating paraboloid, J. Fluid Mech., 17, 257-266, 1963.

Mollo-Christensen, E., Edge waves in a rotating stratified fluid, an exact solution, J. Phys. Oceanogr., 9, 226-229, 1979.

Moore, D. W., Mass transport velocity induced by free oscillations at a single frequency, Geophys. Fluid Dyn., 1, 237-247, 1970.

Moore, D. W., and S. G. H. Philander, Modeling of the tropical oceanic circulation, in Marine Modeling, The Sea, vol.6, edited by E. D. Goldberg, I. N. McCave, J. J. O'Brien, and J. H. Steele, pp. 319-361, John Wiley, New York, 1977.

Moray, R., A relation of some extraordinary tydes in the west-isles of Scotland, Philos. Trans. R. Soc. London, 1, 53-55, 1665.

Munk, W. H., F. E. Snodgrass, and M. Wimbush, Tides offshore: Transition from California coastal to deep-sea waters, Geophys. Fluid Dyn., 1, 161-235, 1970.

Mysak, L. A., On the very low frequency spectrum of the sea level on a continental shelf, J. Geophys. Res., 72, 3043-3047, 1967.

Mysak, L. A., Effects of deep-sea stratification and current on edge waves, J. Mar. Res., 26, 34-42, 1968.

Mysak, L. A., On the generation of double Kelvin waves, J. Fluid Mech., 37, 417-434, 1969.

Mysak, L. A., On the stability of the California undercurrent off Vancouver Island, J. Phys. Oceanogr., 7, 904-917, 1977.

Mysak, L. A., Long-period equatorial topographic waves, J. Phys. Oceanogr., 8, 302-314, 1978a.

Mysak, L. A., Resonant interactions between topographic planetary waves in a continuously stratified fluid, J. Fluid Mech., 84, 769-793, 1978b.

Mysak, L. A., Recent advances in shelf wave dynamics, Rev. Geophys., 18, 211-241, 1980.

Mysak, L. A., Barotropic instability of flow along a trench, Geophys. Astrophys. Fluid Dyn., 19, 1-33, 1982.

Mysak, L. A., and C. L. Tang, Kelvin wave propagation along an irregular coastline, J. Fluid Mech., 64, 241-261, 1974.

Mysak, L. A., P. H. LeBlond, and W. J. Emery, Trench waves, J. Phys. Oceanogr., 9, 1001-1013, 1979.

Mysak, L. A., E. F. Johnson, and W. W. Hsieh, Baroclinic and barotropic instabilities of coastal currents, J. Phys. Oceanogr., 11, 209-230, 1981a.

Mysak, L. A., R. D. Muench, and J. D. Schumacher, Baroclinic instability in a downstream varying channel: Shelikof Strait, Alaska, J. Phys. Oceanogr., 11, 950-969, 1981b.

Niiler, P. P., and L. A. Mysak, Barotropic waves along an eastern continental shelf, Geophys. Fluid Dyn., 2, 273-288, 1971.

Odulo, A. B., and Y. N. Pelinovskiy, Effect of random inhomogeneities of ocean bottom relief on the propagation of Rossby waves, Oceanology, Engl. Transl., 18, 505-507, 1978.

Ou, H. W., On the propagation of free topographic Rossby waves near continental margins, 1, Analytical model for a wedge, J. Phys. Oceanogr., 10, 1051-1060, 1980.

Ou, H. W., and R. C. Beardsley, On the propagation of free topographic Rossby waves near continental margins, 2, Numerical model, J. Phys. Oceanogr., 10, 1323-1339, 1980.

Ou, H. W., and J. R. Bennett, A theory of the mean flow driven by long internal waves in a rotating basin, with application to Lake Kinneret, J. Phys. Oceanogr., 9, 1112-1125, 1979.

Ou, H. W., R. C. Beardsley, D. Mayer, W. C. Boicourt, and B. Butman, An analysis of subtidal current fluctuations in the Middle Atlantic Bight, J. Phys. Oceanogr., 11, 1383-1392, 1981.

Pearson, C. E., Note on Kelvin wave reflection in a channel with an arbitrary end wall, Geophys. Astrophys. Fluid Dyn., 8, 303-309, 1977.

Pedlosky, J., The stability of currents in the atmosphere and ocean, I, J. Atmos. Sci., 21, 201-219, 1964.

Pedlosky, J., The destabilization of shear flow by topography, J. Phys. Oceanogr., 10, 1877-1880, 1980.

Pinsent, H. G., Kelvin wave attenuation along nearly straight boundaries, J. Fluid Mech., 53, 273-286, 1972.

Platzman, G. W., G. A. Curtis, K. S. Hansen, and R. D. Slater, Normal modes of the world ocean, II, Description of modes in the period range 8 to 80 hours, J. Phys. Oceanogr., 11, 579-603, 1981.

Reid, R. O., Effect of Coriolis force on edge waves (i), Investigation of the normal modes, J. Mar. Res., 16, 109-144, 1958.

Rhines, P. B., Slow oscillations in an ocean of varying depth, 1, Abrupt topography, J. Fluid Mech., 37, 161-189, 1969a.

Rhines, P. B., Slow oscillations in an ocean of varying depth, 2, Islands and seamounts, J. Fluid Mech., 37, 191-205, 1969b.

Rhines, P. B., Edge-, bottom-, and Rossby waves, Geophys. Fluid Dyn., 1, 273-302, 1970.

Rhines, P. B., The dynamics of unsteady currents, in Marine Modeling, The Sea, vol. 6, edited by E. D. Goldberg, I. N. McCave, J. J. O'Brien, and J. H. Steele, pp. 189-318, John Wiley, New York, 1977.

Ripa, P., Normal Rossby modes of a closed basin with topography, J. Geophys. Res., 83, 1947-1957, 1978.

Robinson, A. R., Continental shelf waves and the response of sea level to weather systems, J. Geophys. Res., 69, 367-368, 1964.

Romea, R. D., and J. S. Allen, On forced coastal trapped waves at low latitudes in a stratified ocean, J. Mar. Res., 40, 369-401, 1982.

Saint-Guily, B., Oscillations propres dans un bassin tournant de profondeur variable: Modes de seconde classe, in Studi in Onore di Giuseppina Aliverti, pp. 15-25, Instituto Universitario Navale di Napoli, Naples, 1972.

Schott, F., and M. Bock, Determination of energy interaction terms and horizontal wavelengths for low-frequency fluctuations in the Norwegian current, J. Geophys. Res., 85, 4007-4014, 1980.

Simons, T. J., Comments "On the effect of bottom friction on barotropic motion over the continental shelf," J. Phys. Oceanogr., 13, 147-148, 1983.

Smith, P. C., Eddies and coastal interactions, in Eddies in Marine Science, edited by A. R. Robinson, pp. 446-480, Springer-Verlag, New York, 1983.

Smith, R., Non-linear Kelvin and continental-shelf waves, J. Fluid Mech., 52, 379-391, 1972.

Smith, R., Second-order turning point problems in oceanography, Deep Sea Res., 22, 837-852, 1975.

Smith, R., Poleward propagating perturbations in currents and sea levels along the Peru coast, J. Geophys. Res., 83, 6083-6092, 1978.

Suginohara, N., Propagation of coastal trapped waves at low latitudes in a stratified ocean with continental shelf slope, J. Phys. Oceanogr., 11, 1113-1122, 1981.

Suginohara, N., Coastal upwelling: Onshore-offshore circulation, equatorial coastal jet and poleward undercurrent over a continental shelf-slope, J. Phys. Oceanogr., 12, 272-284, 1982.

Taylor, G. I., tidal oscillations in gulfs and rectangular basins, Proc. London Math. Soc., 20, 148-181, 1920.

Thompson, R. O. R. Y., and J. R. Luyten, Evidence for bottom-trapped topographic Rossby waves from single moorings, Deep Sea Res., 23, 629-635, 1976.

Thomson, R. E., On the generation of Kelvin-type waves by atmospheric disturbances, J. Fluid Mech., 42, 657-670, 1970.

Thomson, R. E., and W. R. Crawford, The generation of diurnal period shelf waves by tidal currents, J. Phys. Oceanogr., 12, 635-643, 1982.

Viera, F., and V. T. Buchwald, The response of the East Australian continental shelf to a travelling pressure disturbance, Geophys. Astrophys. Fluid Dyn., 19, 249-265, 1982.

Wang, D.-P., Coastal trapped waves in a baroclinic ocean, J. Phys. Oceanogr., 5, 326-333, 1975.

Wang, D.-P., Low-frequency sea level variability on the Middle Atlantic Bight, J. Mar. Res., 37, 683-697, 1979.

Wang, D.-P., Diffraction of continental shelf waves by irregular alongshore geometry, J. Phys. Oceanogr., 10, 1187-1199, 1980.

Wang, D.-P., Effects of continental slope on the mean shelf circulation, J. Phys. Oceanogr., 12, 1524-1526, 1982.

Wang, D.-P., and C. N. K. Mooers, Coastal-trapped waves in a continuously stratified ocean, J. Phys. Oceanogr., 6, 853-863, 1976.

Wolanski, E., and D. van Senden, Mixing of Burdekin river flood waters in the Great Barrier Reef, Aust. J. Mar. Freshwater Res., 34, 49-63, 1983.

Wright, D. G., On the stability of a fluid with specialized density stratification, 1, Baroclinic instability and constant bottom slope, J. Phys. Oceanogr., 10, 639-666, 1980.

Yamagata, T., A theory for propagation of an oceanic warm front with application to Sagami Bay, Tellus, 32, 73-76, 1980.

# INTERNAL TIDES, INTERNAL WAVES, AND NEAR-INERTIAL MOTIONS

Peter G. Baines

CSIRO Division of Atmospheric Research, Mordialloc, Victoria 3195, Australia

Abstract. The current state of knowledge of inertial oscillations, internal tides, and high-frequency internal waves in continental shelf regions is reviewed. For inertial oscillations, attention is focused on the general effects of a nearby coastline, which are discussed from a theoretical viewpoint, and existing observations are briefly summarized. For internal tides, which are generally less well understood than inertial oscillations, linear generation theory for continental shelves is described, together with a summary of observations. No proper test of the generation theory has so far been made on a shelf. In addition, internal tidal observations show nonlinear character in their generation (producing 6- and 8-hour period waves) and propagation (internal surges and undular bores), and both theory and observations of these aspects are inadequate. High-frequency internal waves are generally observed to propagate shoreward from the deep sea.

## 1. Introduction

In this paper we present a comparison between theoretical knowledge and observations of motion on continental shelves in the inertial, tidal, and higher-frequency bands. These are discussed in order of increasing frequency. Continental shelf regimes vary widely, depending on local weather and climate, tidal amplitudes, shelf width, geography, and so on, and conclusions obtained for one region may not be applicable to another. In particular, low-frequency phenomena such as upwelling and frontal development will have a significant impact on baroclinic motions at the higher frequencies being discussed here. These considerations must be borne in mind in the following presentation, which attempts to distill general conclusions about these motions.

## 2. Inertial Oscillations

In near-surface waters and on continental shelves, inertial oscillations are principally due to forcing by local (or near-local) winds. In the deep ocean the picture is more complex and has been discussed in some detail by Fu [1981].

Models of inertial wave generation by wind forcing in the open ocean have been described by Pollard [1970], and results from a simple model of the mixed-layer motion using observed wind stress have been shown to correlate well with observations [Pollard and Millard, 1970; Kundu, 1976; Pollard, 1980]. The established properties of near-surface deep ocean inertial motion may be summarized as follows: (1) The motion is generated by changes in surface wind stress which occur on time scales much less than the inertial period; successive changes of this nature result in an intermittent character for the inertial wave field. (2) It is predominantly horizontal and approximately circularly polarized with cum sole rotation. (3) It has frequencies slightly higher than f with slow ($\sim$mm s$^{-1}$) downward energy propagation. (4) It is coherent over large horizontal distances.

On continental shelves, one might expect the presence of a nearby coastline to inhibit such motions. However, inertial oscillations of substantial amplitude have been observed within a few kilometers of coastlines, and it is this question of the influence of a coastline upon which we will focus here. Observations of inertial waves on continental shelves in regions reasonably close to coastlines have been reported by Schott [1971a], Kundu [1976], Johnson et al. [1976], Tang [1979], Thomson and Huggett [1981], Millot and Crépon [1981], and Anderson et al. [1983]. The properties of inertial oscillations on shelves generally include those listed above for the "open" ocean. In addition, the amplitude of the motion is observed to decrease as the shoreline is approached, although it remains approximately circularly polarized. This contrasts with the observed character of tidal motion (or, at least, barotropic tidal motion (see next section), where the current ellipses become progressively more elongated alongshore as the shoreline is approached, without substantial amplitude decrease. On the Oregon shelf, for example, inertial current amplitudes up to 20 cm s$^{-1}$ (200 m radius) have been observed only 12 km offshore [Anderson et al., 1983]. This implies substantial gradients in the motion closer to the shoreline.

To investigate this aspect, we consider the following simple two-layer model, with the config-

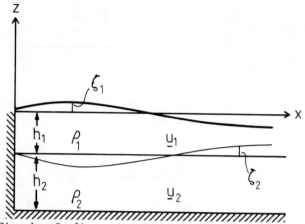

Fig. 1. Configuration for a standard two-layer system.

uration shown in Figure 1. The coastline is situated at $x = 0$, with $y$ alongshore and $z$ vertically upward; $h_j$, $\rho_j$, $u_j$ and $v_j$ denote, respectively, the mean depth, density, and x-(offshore) and y-directed velocity components of the jth layer, where $j = 1$ denotes the upper layer and $j = 2$ the lower; $\zeta_1$ and $\zeta_2$ denote the free surface and interface displacements, respectively. Such two-layer models are applicable in many cases on continental shelves, where the stratification often consists of a surface mixed layer and a bottom mixed layer (maintained by bottom stress), separated by a thermocline. We assume that this system is forced by a wind stress which is transmitted directly to the upper layer and which is independent of the alongshore coordinate $y$. The linear equations governing the motion then are

$$u_{1_t} - fv_1 = -g\,\zeta_{1x} + \tau_{wx}$$

$$u_{2_t} - fv_2 = -g\left(\frac{\rho_1}{\rho_2}\zeta_1 + \frac{\Delta\rho}{\rho_2}\zeta_2\right)_x$$

$$v_{1_t} + fu_1 = \tau_{wy} \qquad v_{2_t} + fu_2 = 0$$

$$(\zeta_1 - \zeta_2)_t + h_1 u_{1x} = 0 \qquad \zeta_{2_t} + h_2 u_{2x} = 0$$

(1)

where $\Delta\rho = \rho_2 - \rho_1$, and $\tau_{wx}$, $\tau_{wy}$ denote the x- and y-directed wind stress components (divided by $\rho_1$), respectively. We will assume that the wind stress has step function structure, initiating suddenly at $t = 0$, so that $\tau_{wx}$, $\tau_{wy}$ are zero for $t < 0$ and constant for $t > 0$.

The model has obvious limitations, in that dissipative processes and vertical propagation are not included, but it is expected to give a good physical picture for several inertial periods after onset of the wind stress and will suffice to demonstrate the effects of the nearby coastline.

The full time-dependent solution to equations

(1) with the sudden onset of the wind stress may be solved by Laplace transforms. Details of the calculation are given by Millot and Crépon [1981]. This solution may be conveniently represented in terms of barotropic (external) and baroclinic (internal) modes, denoted by the suffixes e and i, respectively. In vector notation we have

$$\underset{\sim}{u}_e = \tfrac{1}{2}(\underset{\sim}{u}_1 + \underset{\sim}{u}_2) \qquad \underset{\sim}{u}_i = \tfrac{1}{2}(\underset{\sim}{u}_1 - \underset{\sim}{u}_2)$$

$$\underset{\sim}{u}_1 = \underset{\sim}{u}_e + \underset{\sim}{u}_i \qquad \underset{\sim}{u}_2 = \underset{\sim}{u}_e - \underset{\sim}{u}_i$$

(2)

and the solution is given by

$$u_e = \frac{h_1}{H} U\left(\frac{x}{\lambda_e},\ ft\right) \qquad u_i = \tfrac{1}{2} U\left(\frac{x}{\lambda_i},\ ft\right)$$

$$v_e = \frac{\tau_{wy}\, t}{2} - \frac{h_1}{H}\int_0^{ft} U\left(\frac{x}{\lambda_e},\ \tau\right)d\tau$$

$$v_i = \frac{\tau_{wy}\, t}{2} - \frac{1}{2}\int_0^{ft} U\left(\frac{x}{\lambda_i},\ \tau\right)d\tau$$

(3)

$$\zeta_2 = -\frac{h_2}{f}\int_0^{ft}\frac{\partial}{\partial x}\left[\frac{h_1}{H}U\left(\frac{x}{\lambda_e},\ \tau\right) - \frac{1}{2}U\left(\frac{x}{\lambda_i},\ \tau\right)\right]d\tau$$

where

$$U(X,\tau) = \frac{\tau_{wx}}{f}F(X,\ \tau) + \frac{\tau_{wy}}{f}\int_0^{\tau}F(X,\ \tau')\,d\tau'$$

(4)

$$\frac{dF}{dX}(x,\tau) = 0 \qquad\qquad 0 < \tau < X$$

$$= J_0\left[(\tau^2 - X^2)^{1/2}\right] \quad 0 < X < \tau$$

(5)

Here $H = h_1 + h_2$ is the total depth, and $\lambda_e$, $\lambda_i$ are the respective barotropic and baroclinic radii of deformation, given (to a very good approximation) by

$$\lambda_e = \frac{\sqrt{gH}}{f} \qquad \lambda_i = \left(\frac{g\Delta\rho\, h_1 h_2}{\rho_2 H}\right)^{1/2}\Big/ f$$

(6)

$J_0$ denotes the Bessel function of zero order. For large times, asymptotic expressions may be readily found for the various integrals in equations (3) and (4). For example,

$$F(X,\ \tau) = \sin\tau \qquad 0 < \tau \leqslant X$$

(7)

$$F(X,\ \tau) = X\left\{\frac{2}{\pi\tau}\right\}^{1/2}\cos\left(\tau - \frac{\pi}{4}\right)\left[1 + 0\left(\frac{1 + X^2}{\tau}\right)\right]$$

$$\frac{1 + X^2}{\tau} \ll 1$$

The barotropic component of the above solution has been given by Crepon [1967] and is described by Csanady [1982]. For the present two-layer system, we note that the barotropic and baroclinic parts of the solution for the velocity fields are very similar, the only differences being in the amplitude and in the length scales, $\lambda_e$ and $\lambda_i$. The barotropic or baroclinic motions dominate the flow depending on whether $h_1/H$ is greater or less than $1/2$, respectively. For each mode the nature of the solution is as follows. At each point the motion initially has the "open ocean" behavior, with net transport to the right (for $f > 0$) of the wind stress and inertial oscillations superimposed. For $t > 0$, however, the effect of the coast propagates seaward at the appropriate wave speed ($f\lambda_e$ or $f\lambda_i$), and after the passage of this "front" the motion consists of three parts: (1) a residual geostrophic longshore flow with $e^{-x/\lambda}$ structure, (2) the open ocean Ekman transport with $1 - e^{-x/\lambda}$ structure, and (3) decaying $[0(1/ft)^{1/2}]$ inertial oscillations. The dispersion relation for these propagating oscillations is

$$\omega^2 = f^2(1 + \lambda^2 k^2)$$

where $k$ is the wave number, so that they have frequencies slightly in excess of $f$.

We take typical parameter values to be

$$H = 50 \text{ m} \qquad h_1 = h_2 = 25 \text{ m}$$
$$f = 10^{-4} \text{ rad}^{-1} \qquad \Delta\rho = 10^{-3} \text{ g cm}^{-3} \qquad (8)$$

and obtain

$$\lambda_e \doteq 220 \text{ km} \qquad \lambda_i = 3.5 \text{ km} \qquad (9)$$

For regions adjacent to the coastline we will have $x/\lambda_e \leqslant 1$, and for these regions we may identify three main "regimes" depending on the elapsed period of time after the wind change, as follows.

1. $ft < x/\lambda_e$. The motion consists of the undisturbed open ocean flow of constant transport directed to the right of the wind stress plus inertial oscillations in the upper layer; there are no pressure gradients and no motion in the lower layer. The solution is

$$\underset{\sim}{u}_1 = \frac{\tau_w}{f} \sin ft - \frac{\hat{z} \wedge \tau_w}{f}(1 - \cos ft)$$
$$\underset{\sim}{u}_2 = 0 \qquad \zeta_1 = \zeta_2 = 0 \qquad 0 < ft < \frac{x}{\lambda_e} \qquad (10)$$

where $\tau_w$ is the stress vector and $\hat{z}$ is the unit vector in the vertical. Near the coast this regime will be short-lived.

2. $x/\lambda_e < ft < x/\lambda_i$. The barotropic wave from the coast has passed, so that the barotropic component of the oscillations of equations (10) is decaying as $0(1/ft)^{1/2}$. Inertial oscillations are now present in the lower layer driven by the barotropic pressure gradient and decaying in the same manner after the initial onset. There are no

significant inertial oscillations in $\zeta_2$. If $\lambda_i/\lambda_e \ll 1$, a stage will be reached where the inertial oscillations are predominantly baroclinic.

3. $ft > x/\lambda_i$. The baroclinic wave front has now also passed the point $x$, and the baroclinic oscillations are decaying as $0(1/ft)^{1/2}$. Immediately after the passage of the baroclinic wave front, significant inertial oscillations will be present in the interface displacement $\zeta_2$. When the inertial oscillations decay, the remaining motion has the well-known form

$$u_1 = \frac{\tau_{wy}}{f}\left[1 - \frac{h_2}{H}e^{-x/\lambda_i} - \frac{h_1}{H}e^{-x/\lambda_e}\right]$$

$$u_2 = \frac{\tau_{wy}}{f}\frac{h_1}{H}\left[e^{-x/\lambda_i} - e^{-x/\lambda_e}\right]$$

$$v_1 = \frac{\tau_{wx}}{f}\left[1 - \frac{h_2}{H}e^{-x/\lambda_i} - \frac{h_1}{H}e^{-x/\lambda_e}\right]$$
$$+ \int_0^t \tau_{wy}\,dt\left[\frac{h_2}{H}e^{-x\lambda_i} + \frac{h_1}{H}e^{-x/\lambda_e}\right]$$

$$v_2 = -\left[\frac{\tau_{wx}}{f} + \int_0^t \tau_{wy}dt\right]\frac{h_1}{H}\left[e^{-x/\lambda_i} - e^{-x/\lambda_e}\right] \quad (11)$$

$$\zeta_1 = -\left[\frac{\tau_{wx}}{f} + \int_0^t \tau_{wy}\,dt\right]\frac{h_1}{\sqrt{gH}}e^{-x/\lambda_e}$$

$$\zeta_2 = \left[\frac{\tau_{wx}}{f} + \int_0^t \tau_{wy}dt\right]\left[\left(\frac{h_1 h_2}{(\Delta\rho/\rho_2)gH}\right)^{1/2}e^{-x/\lambda_i} - \frac{h_1 h_2}{\sqrt{gH}\,H}e^{-x/\lambda_e}\right]$$

Expressed in this form, solution (11) is the same as that which would be obtained if the wind stress were increased very slowly (quasi-statically) to its present value, so that no inertial oscillations would be generated at all.

In near-coastal regions, $\lambda_i$ may be quite small ($<3$ km) because of shallow depth and weak stratification, so that regime (2) above may be quite common. Regime (3) should also be significant, and this mechanism is a plausible candidate to explain temperature oscillations of inertial period observed on continental shelves. Schott [1971a] observed baroclinic inertial oscillations in the North Sea which could possibly be interpreted as flow in regime 3, although in practical situations, horizontal variations in wind stress complicate the picture. Millot and Crépon [1981] described observations made during several summers in the Gulf of Lions, which showed that the two-layer model is applicable in this region. In particular, they made the following observa-

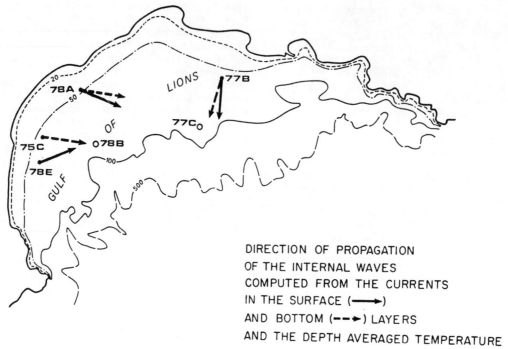

DIRECTION OF PROPAGATION
OF THE INTERNAL WAVES
COMPUTED FROM THE CURRENTS
IN THE SURFACE (———►)
AND BOTTOM (––►) LAYERS
AND THE DEPTH AVERAGED TEMPERATURE

Fig. 2. The Gulf of Lions, Mediterranean Sea (center 43°N, 4°E, approximately). Directions of propagation of internal waves at (or slightly above) the inertial frequency, obtained from coherence computations between the depth-averaged temperature and the surface or bottom currents. These quantities were largely incoherent at locations 77C and 78B, presumably due to superposition of waves from different directions [from Millot and Crépon, 1981].

tions: (1) Motion in the upper layer had a peak in the energy spectrum very close to the inertial frequency, whereas the spectra of motion in the lower layer and the temperature field had broader peaks at frequencies a little larger than inertial, indicating that these disturbances were due to propagating waves. (2) In this same band of frequencies, surface and bottom currents at the same mooring were coherent with a phase lag of π. (3) At distances from shore less than 20 km the temperature fluctuations near inertial frequency were coherent with the currents. Points 2 and 3 indicate that the flow is predominantly in regime 3. With this assumption, Millot and Crépon [1981] deduced the directions of propagation of the internal waves based on the required phase relationship between the temperature variations and the currents in each of the layers, and these are shown in Figure 2. That these are directed away from the nearby coastline is strong vindication for the model. Further out in the Gulf the same currents and temperatures are incoherent, due to superposition of waves from different parts of the coast.

This model may be extended by the inclusion of other factors (in particular, the history of wind stress change may be included by use of a convolution integral), but its purpose here is to identi-

fy coastline effects and to describe a possible basis for interpretation of observations. Regions where the stratification differs substantially from two-layer will probably require the inclusion of more vertical modes. Kundu et al. [1983] have considered the complementary system of a mixed layer above a region of constant density gradient, with viscosity and diffusion; their more detailed results parallel those for the two-layer model described here.

The above analytic solution may be applied to the interpretation of other situations with different initial conditions, such as those shown in Figure 3. In Figures 3a and 3b, a longshore current is assumed to give geostrophic balance. Since the flow is hydrostatic in this model, the solutions for a uniform wind stress with these mean density fields may be obtained from equations (2)–(4) but with the origin for the barotropic motion at $x = 0$ and that for the baroclinic motion at $x = x_1$. Ultimately, of course, a persistent wind stress will result in a change in the mean state, particularly the interface $\zeta_2$ (see equation (11)). This limits the applicability of the model, but for inertial oscillations the important parameter will be the value of $x_1$ when the wind change occurs.

The major significance of inertial oscillations

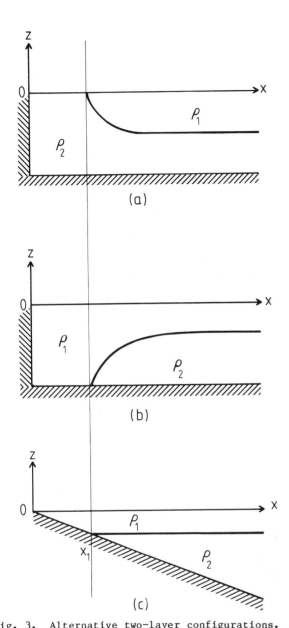

Fig. 3. Alternative two-layer configurations.

# 3. Internal Tides

Internal tides are generated by the advection of stratified water across bottom contours by the barotropic tide. General reviews of the phenomenon have been given by Wunsch [1975] and Schott [1977], and some aspects of coastal tidal currents have been reviewed by Winant [1979]. Models for internal tide generation were pioneered by Rattray [1960] and Rattray et al. [1969]. For continental shelf regions the generation process is generally strongest at or near the shelf break, where the generating force is largest [Baines, 1973]. On the shelf (depth $\leq 100$ m), baroclinic tidal energy is frequently observed to be concentrated in the lowest vertical mode, although Torgrimson and Hickey [1979] have observed a beam structure (implying many modes) on the Oregon shelf. The present state of theoretical models for internal tide generation has been described by Baines [1982], and these linear models give good quantitative predictions in laboratory experiments for continuous stratification, but nonlinear effects readily become significant in layered stratification [Baines and Fang, 1985]. In section 7 of Baines [1982] a simple procedure is given for the calculation of internal tides generated from a simple shelf/slope topography, with a mean stratification consisting of a surface mixed layer surmounting a shallow thermocline and a deeper stratified region. In this situation the wave amplitudes and velocity fields may be extracted from graphs (i.e., no computation is required), so that a direct comparison with oceanographic data may readily be made. To emphasize this point, we reproduce below the relevant equations for the motion on the continental shelf.

Continental slope. The number of detailed observations of internal tides over shelves and slopes is not large. Data over the continental slope described by Wunsch and Hendry [1972], Barbee et al. [1975], Petrie [1975], Horn and Meincke [1976], Schott [1977], Torgrimson and Hickey [1979], and Gordon [1979] are qualitatively consistent with the expected theoretical beam structure [Rattray et al., 1969; Baines, 1974, 1982; Prinsenberg et al., 1974] without confirming it. A quantitative comparison between model predictions and observations at two locations at 500 m depth on the northwest African shelf by Huthnance and Baines [1982] gave poor results. The reasons for this are felt to be twofold. First, the observed internal tide motion is intermittent (as it is in most places), and this is attributed to varying stratification in the shelf break region, where the generating body force is concentrated. The shelf break region is relatively close to the surface, and the stratification may vary on long time scales because of a number of processes [see Huthnance, 1981], particularly in upwelling regions such as northwest Africa. The effects of varying stratification between the generation region and the observation point may also

in coastal processes may well be their influence on vertical mixing and entrainment. They produce large vertical shears, and consequently low Richardson numbers, much more frequently than in situations where they are absent, because the shear rotates and at some phase will add to that of other motions. Confident estimates of the magnitude of this effect are not yet available.

Another possible source of inertial oscillations is small-scale geostrophic adjustment in coastal currents caused by factors other than the wind, such as river surges. This has been discussed by Tang [1979], but convincing observations have not yet been made.

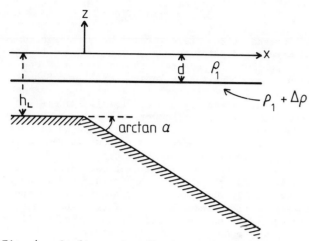

Fig. 4. Configuration for internal tide generation model.

thermocline of magnitude $\Delta\rho$ and a continuously stratified layer with Brunt-Väisälä frequency N. For the latter we may treat generation first on the interface (thermocline) and then subsequently on the stratified continuum if (approximately)

$$R > 1/(1 + \frac{\pi}{2}\sqrt{S}) \qquad (12)$$

where

$$R = \frac{d}{h_L} \qquad S = \frac{g\,\Delta\rho}{\rho dN^2} \qquad (13)$$

Conditions for the applicability of these models to particular oceanic stratification are discussed in Section 13 of Baines [1982]. For the case N = const, with steep topography ($\alpha > c$, where c is the gradient of the internal wave rays), the internal tides on the shelf are largely associated with singularities in the linear solution; consequently, they will have short wavelengths and be dissipated quickly. In practice, the linear model is not accurate for these details, and in laboratory experiments, amplitudes on the continental shelf are very small in these situations [Baines and Fang, 1983]. For flat topography ($\alpha < c$) the internal tide amplitudes are usually not large unless $\alpha/c$ is close to unity, in which case the foregoing remarks apply also.

For the second case, where the stratification is largely concentrated into a narrow thermocline, from Section 7 of Baines [1982] the tidal velocities on the shelf may be written as

$$\underset{\sim}{u} = \underset{\sim}{u}_1 + \underset{\sim}{u}_A + \underset{\sim}{u}_B \qquad (14)$$

where $\underset{\sim}{u}_1$ denotes the barotropic tide and

$$u_A = \frac{QD_0}{d} e^{-i(\theta+\omega t)} \qquad\qquad -d < z < 0$$

$$u_A = \frac{-QD_0}{h_L - d} e^{-i(\theta+\omega t)} \qquad -h_L < z < -d \qquad (15)$$

$$\zeta_0 = \frac{Q}{L\omega}\frac{D_0}{(1-R)^{1/2}} e^{-i(\theta+\omega t)} \qquad v_A = -\frac{if}{\omega} u_A$$

where the real part is taken, $\zeta_0$ is the interface displacement, and

$$\theta = \frac{x}{L(1-R)^{1/2}} \qquad L^2 = \frac{g\Delta\rho d}{\rho(\omega^2-f^2)} \qquad (16)$$

$\omega$ is the tidal frequency and Q is given by

$$h_L u_1 = Q\cos\omega t \qquad (17)$$

at the shelf break (x = 0). $D_0$ is given by

be a factor, as this changes the ray geometry [Hayes and Halpern, 1976]. Second, the model assumes two-dimensional (y-independent) coastline and topography. Barotropic tidal ellipses become elongated alongshore as the coastline is approached, so that longshore topographic variations become much more effective generators of internal tides than cross-shore variations, particularly for narrow shelves. Such variations may result in internal Kelvin waves, as evidently observed by Huthnance and Baines [1982]. The inclusion of longshore topographic variations is probably the next major step in the development of internal tide generation theory.

In some locations, baroclinic tidal energy at the $S_2$ frequency has been found to be larger than for $M_2$, although barotropic $S_2$ energy is substantially less than for $M_2$ [Gould and McKee, 1973; Wolanski and Pickard, 1983]. The most plausible explanation for this initially surprising result is that the barotropic current ellipse for $S_2$ is oriented more perpendicularly across the topographic contours than for $M_2$, in the generation region for the observed waves, but this is as yet unconfirmed.

Double-frequency (~6-hour period) baroclinic tides are frequently observed over continental slopes [e.g., Petrie, 1975; Gordon, 1979; Huthnance and Baines, 1982]; these probably originate from nonlinear processes near the shelf break, but the detailed mechanics are not clear.

Continental shelf. We first consider the results from theoretical models of generation of internal tides at the shelf break and then discuss the observations. We take the idealized geometry as shown in Figure 4 and assume that there is no reflection from the coastline; $h_L$ is the fluid depth on the shelf, and $\alpha$ is the slope of the continental slope. We consider two types of stratification: (1) N-constant and (2) a surface mixed layer of thickness d surmounting a shallow

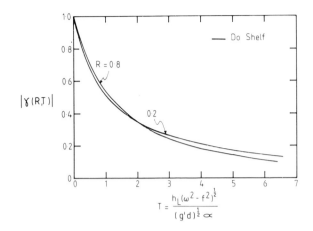

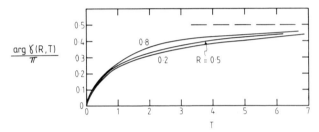

Fig. 5. The amplitude function γ(R, T) for motion on the continental shelf [from Baines, 1982].

$$D_0 = - \frac{R}{1 + (1 - R)^{-1/2}} \gamma(R, T)$$

where

$$T = \frac{h_L}{L\alpha} = \frac{h_L(\omega^2 - f^2)^{1/2}}{(g\Delta\rho/\rho)^{1/2}\alpha} \qquad (18)$$

and γ(R, T) is given in Figure 5. $\underset{\sim}{u}_B$ is generated in the lower stratified layer and is generally negligible on the shelf for the same reasons as those which apply to case (1). The resulting internal tide is an interfacial wave propagating toward the coast with a typical wavelength of 20-30 km. From this linear picture we may infer that internal tides propagating shoreward on the shelf may be substantial for (approximately) two-layer stratification but may be negligible for near-uniform stratification (although for complex shelf/slope topography this generalization may not hold [e.g., Torgrimson and Hickey, 1979]). Here it is the stratification near the shelf break which counts; as it propagates onshore, a wave will evolve with little energy loss if the stratification evolves from two-layer-type to constant-N-type.

The model therefore gives us a picture of long linear internal waves propagating shoreward but sensitive to the stratification near the shelf break. How does this picture match up with observations? Continental shelves vary widely in their stratification properties, and the observed internal tide varies substantially in character from shelf to shelf. It may also vary with season: generally speaking, in the tropics, water on continental shelves remains stratified throughout the year, but in mid-latitudes the water is stratified in spring and summer but tends to be well mixed (or weakly stratified) in winter, depending on the combined effects of surface cooling and wind and tidal stirring.

The observations to date are summarized in Table 1. All these observations are of semidiurnal tides. In most cases, insufficient data are available to enable a comparison with the model described above. In locations where internal tides are not observed, this may be due to the presence of constant N (approximately) conditions near the shelf break, so that the waves are not generated; alternatively, if the density structure on the shelf is of the type shown in Figure 3a or 3b, the waves may be reflected or absorbed by the frontal structure.

Another imponderable factor is the reflection coefficient from the shoreline. For constant N stratification the linear theory predicts that the waves disappear into the "corner regions" (along reflected characteristics) where they are dissipated, and this has been qualitatively verified by Cacchione and Wunsch [1974]. For two-layer-type stratification (e.g., Figure 3c) the reflection coefficient depends on the bottom slope and, probably, the wave amplitude and wavelength and the mixed-layer depth, and it is largely unknown. The only known laboratory observations of reflection coefficients of internal waves incident on a beach in a two-layer system are those of Nagashima [1971], and these are not directly applicable to the ocean (in particular, the density ratio (upper layer/lower layer) is not specified but is probably about 0.8). The reflection coefficient for the observations of Winant and Bratkovich [1981] appears to be approximately unity, and this may be due to the fact that the wavelength is much larger than the shelf width. This suggests that for narrow shelves the "perfect reflection" model of Rattray [1960] may be appropriate.

In some cases, most notably the Landsat observations, the internal tide takes the form of periodic surges (~12 hours apart) with a number of associated smaller-scale (~100-700 m length) waves. These may be plausibly attributed to the steepening of the linear near-interfacial waves described by equations (15) to form undular bores. This may be expected when the upper layer is thinner than the lower layer [e.g., Maxworthy, 1979] or vice versa, but this process has yet to be satisfactorily described and substantiated in the ocean (see next section), although the bores themselves have been described [e.g., Cairns, 1967].

Some observations [e.g., Jones and Padman,

TABLE 1. Observations of Internal

| Author | Location | Position | Water Depth | Stratification |
|--------|----------|----------|-------------|----------------|
| Schott [1971b] | North Sea | 56°20'N, 1°E | 82 m | ~ two layer |
| Gordon [1978] | NW Africa | 21°40'N | 65 m, 100 m | variable |
| Huthnance and Baines [1982] | NW Africa | 22°44'N | 74 m | variable |
| Petrie [1975] | Scotian shelf | 43°N, 63°W | 100–135 m | variable |
| Halpern [1971] | Massachusetts Bay | 42°16'N, 70°25'W | 82 m | variable |
| Haury et al. [1979] | Massachusetts Bay | as above (approximately) | various | variable |
| Chereskin [1983] | Massachusetts Bay | as above (approximately) | various | variable |
| Sawyer and Apel [1976] | NE America | 36°N–44°N | various | variable |
| Apel et al. [1975] | SW Africa | 30°S–32°S | various | variable |
| Hayes and Halpern [1976], Wang and Mooers [1977] | Oregon shelf | 45°15'N | 100 m | variable |
| Torgrimson and Hickey [1979] | Oregon shelf | 45°16N | 50, 100 m | ~ constant N but variable |
| Summers and Emery [1963] | S California shelf | 31°30'N–34°30'N | >50 m | variable |
| Cairns [1967] | S California shelf | 32°45'N | 18 m | variable |
| Winant and Bratkovich [1981] | S California shelf | 32°57'N | 15, 30, 60 m | ~ constant N, but variable, mean varies with season |
| Apel and Gonzalez [1983] | Baja California shelf | 25–26°N | various | variable |

Tides on Continental Shelves

| Propagation Direction | Type of Observation | Time of Year | Comments |
|---|---|---|---|
| northward | C, T | Sept. | Waves presumably generated at Dogger Bank or other topography in southern North Sea |
| onshore | C, T | March–April | |
| | C, T | Feb. | No significant baroclinic tides observed |
| | C, T | various spanning most months | No significant baroclinic tides observed |
| onshore (westward) | T | July, Aug. | Internal surge with undular bore structure |
| eastward and westward | C, T, acoustic | Aug. | Internal surge with undular bore structure |
| eastward and westward | C, T, acoustic | Sept. | Internal surge with undular bore structure |
| onshore | Landsat | July, Aug. | Many photographs, showing bands of smaller-scale waves to be very common |
| onshore | Landsat | Nov. | |
| shoreward | C, T | July | |
| onshore | C, T | July, Aug. | Beam structure identified |
| onshore | T | Aug., March/April | Multiple ship survey, waves observed propagating on-shore from deep water |
| onshore | T | June–Aug. | Asymmetric surges with vertical temperature faces propagating shoreward |
| | C, T | all months except April | Standing wave structure |
| onshore | Seasat SAR | July | Many wave packets |

TABLE 1.

| Author | Location | Position | Water Depth | Stratification |
|---|---|---|---|---|
| Jones and Padman [1983] | Bass Strait | 38°26'S, 148°12'E | 78 m | ~ two layer |
| Baines [1981] | Australian NW shelf | 12°S–22°S | various | variable |
| Holloway [1983] | Australian NW shelf | 19°30'S–20°30'S | various | strong but variable |
| Holloway [1985] | Australian NW shelf | 20°S–21°30'S | 80–150 m | ~ constant N but variable |

*C denotes current, T temperature.

1983] show the presence of 6-hour (frequency = 2 x semidiurnal] and 8-hour (frequency = diurnal + semidiurnal) period waves on the shelf, and Schott [1971b] has observed 3-hour (167-min) period waves in the North Sea. As with the flow over the slope, this involves nonlinear generation mechanisms near the shelf-break (or other topography) which are not understood.

There is scope for the compilation of a worldwide climatology of internal tides on continental shelves, but the data of Table 1 are too sparse for such a task. Theoretical estimates for all continental shelves have been compiled by the author, and the magnitudes for the most significant regions are given by Baines [1982].

### 4.  Internal Waves

It has been known for some time that internal waves on the continental shelf tend to propagate toward the shoreline [La Fond, 1962; Cox, 1962].

Studies of internal waves on shelves have mostly been focused on the larger-amplitude wave trains which are of tidal origin and are generated at the shelf break or other topographic features such as ridges and banks. These studies have been included in Table 1; another striking example in the Straits of Georgia has been described by Gargett [1976]. The ubiquity of these waves has been shown via numerous Landsat observations [e.g., Apel et al., 1975]; they seem to be common where the water column is stratified and onshore tidal currents are large at the shelf break (or relevant topography). Wavelengths are typically of the order of several hundred meters. Sawyer [1983] has shown that the origin of these wave trains

generally corresponds to the flood phase of the tide at the shelf break. Subsurface observations of these phenomena have been made in fjords, but the oceanic region where they have received the most attention is Massachusetts Bay [Halpern, 1971; Haury et al., 1979; Chereskin, 1983]. Here the waves are apparently generated in both directions (onshore and offshore) by tidal flow over an asymmetric ridge (Stellwagen Bank). Two generation mechanisms for this type of wave train have been proposed: the waves may be generated as a lee wave train which propagates over the topography after the tide turns, or they may result from a single large-amplitude lee wave which evolves to anundular bore structure after the tide turns. The details seem to depend on the bottom shape [Chereskin, 1983].

Wave trains and undular bores with similar character but generally smaller amplitude may be generated by wind effects. Such wave trains have been observed in lakes and also on the Caspian Sea shelf [Ivanov et al., 1981].

The only quantative study of the offshore variability of the whole band of high-frequency internal waves [tidal frequency < ω < N) is that of Gordon [1978] on the JOINT-I data off the northwest coast of Africa (latitude 21°40'N). Data were obtained at seven moorings along a line approximately perpendicular to the bottom contours, from depths ranging from 1200 m to 40 m in March and April. The general conclusions were as follows:

1. For depths less than 800 m, a substantial fraction (25–45%) of the internal wave energy was in (vertical) mode 1, propagating shoreward. Seaward of this region there was a transition

Continued

| Propagation Direction | Type of Observation | Time of Year | Comments |
|---|---|---|---|
| not known | T | Nov.– March | In addition to the $M_2$ period, significant motion with 8.2-hour and 6.25-hour periods were present |
| onshore | Landsat | Oct.– Nov. | Many wave packets in some areas |
| onshore | C | Jan.–May | Large-amplitude internal tides observed near the shelf break, which appear to dissipate rapidly onshore |
| | C, T | all months except Sept., Oct. | Bottom currents intensified when bottom slope close to ray slope |

region where the internal wave field merged into the isotropic deep ocean spectrum.

2. The energy source for this wave field was apparently the deep ocean, rather than (for example) the shelf break region.

3. As the waves propagated shoreward, the energy density increased until a saturation level was reached, near the shelf break; shoreward of this point the wave amplitudes and energies were saturated and were controlled by the stratification and the total depth h. The energy flux and energy density are shown in Figure 6. Wave energy dissipation occurred on the shelf and was largest near the shelf break; there was no significant reflection back from the coastline. Similar on-shore propagation has been reported by Winant and Bratkovich [1981] on the Southern California shelf.

This picture parallels the familiar picture for surface waves; waves are generated in the deep sea and when close to shore are refracted toward it by the shallowing depth, until they reach a surf zone where they break and are dissipated. For the internal waves the "surf zone" is the entire continental shelf. Distant storms, etc., (though not so distant as for surface waves) may be expected to result in increased incident energy fluxes over the continental slope, although Gordon [1978] found no correlation between the internal wave field and local winds, tides, or currents.

It is tempting to apply this model to most continental shelves; it is probably valid in general, although it is obviously not the whole story. As discussed above, tidal generation near the shelf break results in substantial high-frequency wave generation in many locations, and the

effect of this on the internal wave spectrum as a whole is not clear.

5. Summary and Discussion

The foregoing account of motion in the three frequency bands indicates that there are still substantial gaps in our understanding, particularly for internal waves and tides, as evidenced by the inadequacy of theoretical models to explain significant features of the observations.

For inertial oscillations the effect of a nearby coastline has been shown to be consistent with predictions from a linear two-layer model (involving baroclinic current and temperature oscillations), but the effects of inertial oscillations on other processes such as vertical mixing have yet to be determined. The effect of the bottom boundary layer on downward propagating inertial oscillations on continental shelves is another conspicuous area where our understanding is incomplete.

Internal tides should, in principle, be more predictable than inertial oscillations, but in fact the situation is worse. Simple models for waves on the shelf, generated at the shelf break, have yet to be tested in detail. Nonlinear generation processes resulting in 6-hour and 8-hour period waves are not uncommon; nor are nonlinear propagation effects giving internal undular bores and surges. Neither can be satisfactorily modeled at present. The notorious intermittency of internal tides is most probably due to variable stratification in the generation region and elsewhere, but this has yet to be verified. Reflection coefficients from the shoreline seem to be

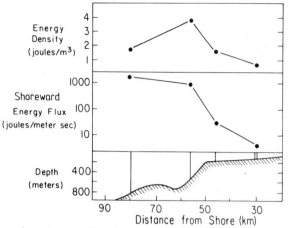

Fig. 6. Energy density and energy flux in the first vertical mode, integrated over the frequency range 0.2–2 cph (approximately) [from Gordon, 1978].

large in some cases and need to be determined in general. Three-dimensional topographic features such as submarine canyons probably have a substantial effect on tidal and internal wave processes on the shelf [Hotchkiss and Wunsch, 1982; Baines, 1983]. From the purely observational point of view, compilation of a climatology of internal tides on continental shelves should be quite feasible, but the data are currently far too sparse.

The higher-frequency internal waves are less important and seem less mysterious as regards their propagation on shelves and slopes (although observations are few), but the overall contribution of the tides to these motions is obscure.

The most significant aspects of all three frequency bands for applications and long-period motion generally are probably as follows:

1. Their contributions to vertical mixing when the fluid is density stratified.

2. Contributions to horizontal diffusion of passive contaminants.

3. Rectified currents resulting from aspect 1. For example, mixing at the shelf break or shoreline may result in horizontal density gradients and resultant geostrophic flows. These effects are not easy to estimate; McEwan [1983] has shown that breaking internal waves may convert 25% of their energy to potential energy of the stratification through mixing; if this ratio is widely applicable to the ocean, these effects will not be insignificant.

4. In some locations, baroclinic (and barotropic) tidal currents may be sufficiently large to dominate the motion; on parts of the Australian northwest shelf, for example, baroclinic tidal currents of 150 cm/s have been recorded.

### References

Anderson, I., A. Huyer, and R. L. Smith, Near-inertial motions off the Oregon coast, J. Geophys. Res., 88, 5960-5972, 1983.

Apel, J. R., and F. I. Gonzalez, Nonlinear features of internal waves off Baja California as observed from the Seasat imaging radar, J. Geophys. Res., 88, 4459-4466, 1983.

Apel, J. R., H. M. Byrne, J.R. Proni, and R. L. Charnell, Observations of oceanic internal and surface waves from the Earth Resources Technology Satellite, J. Geophys. Res., 80, 865-881, 1975.

Baines, P. G., The generation of internal tides by flat-bump topography, Deep Sea Res., 20, 179-205, 1973.

Baines, P. G., The generation of internal tides over steep continental slopes, Philos. Trans. R. Soc. London, Ser. A, 277, 27-58, 1974.

Baines, P. G., Satellite observations of internal waves on the Australian north-west shelf, Aust. J. Mar. Freshwater Res., 32, 457-63, 1981.

Baines, P. G., On internal tide generation models, Deep Sea Res., 29, 307-338, 1982.

Baines, P. G., Tidal motion in submarine canyons-- A laboratory experiment, J. Phys. Oceanogr., 13, 310-328, 1983.

Baines, P. G., and X-H. Fang, Internal tide generation at a continental slope: A comparison between theory and experiment, Dyn. Atmos. Oceans, 9, 297-314, 1985.

Barbee, W. B., J. G. Dworski, J. D. Irish, L. H. Larsen, and M. Rattray, Measurement of internal waves of tidal frequency near a continental boundary, J. Geophys. Res., 80, 1965-1974, 1975.

Cacchione, D., and C. Wunsch, Experimental study of internal waves over a slope, J. Fluid Mech., 66, 223-239, 1974.

Cairns, J. L., Asymmetry of internal tidal waves in shallow coastal waters, J. Geophys. Res., 72, 3563-3565, 1967.

Chereskin, T. K., Generation of internal waves in Massachusetts Bay, J. Geophys. Res., 88, 2649-2661, 1983.

Cox, C. S., Internal waves, II, in The Sea, vol. I, edited by M. N. Hill, pp. 752-763, Interscience, New York, 1962.

Crépon, M., Hydrodynamique marine en regime impulsionnel, II, Cah. Oceanogr., 19, 847-880, 1967.

Csanady, G. T., Circulation in the Coastal Ocean, 279 pp., D. Reidel, Hingham, Mass., 1982.

Fu, L.-L., Observations and models of inertial waves in the deep ocean, Rev. Geophys., 19, 141-170, 1981.

Gargett, A. E., Generation of internal waves in the Strait of Georgia, British Columbia, Deep Sea Res., 23, 17-32, 1976.

Gordon, R. L., Internal wave climate near the coast of northwest Africa during JOINT-I, Deep Sea Res., 25, 625-643, 1978.

Gordon, R. L., Tidal interactions in a region of large bottom slope near northwest Africa during JOINT-I, Deep Sea Res., 26A, 199-210, 1979.

Gould, W. J., and W. D. McKee, Vertical structure of semi-diurnal tidal currents in the Bay of Biscay, Nature, 244, 88-91, 1973.

Halpern, D., Semidiurnal internal tides in Massachusetts Bay, J.Geophys. Res.,76,6573-6584,1971.

Haury, L. R., M. G. Briscoe, and M. H. Orr, Tidally generated internal wave packets in Massachusetts Bay, Nature, 278, 312-317, 1979.

Hayes, S. P., and D. Halpern, Observations of internal waves and coastal upwelling off the Oregon coast, J. Mar. Res., 34, 247-267, 1976.

Holloway, P. E., Internal tides on the Australian north-west shelf: A preliminary investigation, J. Phys. Oceanogr., 13, 1357-1370, 1983.

Holloway, P. E., A comparison of semi-diurnal internal tides from different bathymetric locations on the Australian north-west shelf, J. Phys. Oceanogr., 15, 240-251, 1985.

Horn, W., and J. Meincke, Note on the tidal current field in the continental slope area off north-west Africa, Mem. Soc. R. Sci. Liege, 6, 31-42, 1976.

Hotchkiss, F. L., and C. Wunsch, Internal waves in Hudson canyon with possible geological applications, Deep Sea Res., 29, 415-442, 1982.

Huthnance, J., Waves and currents near the continental shelf edge, Prog. Oceanogr., 10, 193-226, 1981.

Huthnance, J., and P. G. Baines, Tidal currents in the northwest African upwelling region, Deep Sea Res., 29, 285-306, 1982.

Ivanov, V. A., K. V. Konyayev, and A. N. Serebryanyy, Groups of intense internal waves on the sea shelf, Izv. Acad. Sci. USSR Atmos. Oceanic Phys., Engl. Transl., 17, 966-972, 1981.

Johnson, W. L., J. C. Van Leer, and C. N. K. Mooers, A cyclesonde view of coastal upwelling, J. Phys. Oceanogr., 6, 556-574, 1976.

Jones, I. S. F., and L. Padman, Semi-diurnal internal tides in eastern Bass Strait, Aust. J. Mar. Freshwater Res., 34, 143-153, 1983.

Kundu, P. K., An analysis of inertial oscillations observed near Oregon coast, J. Phys. Oceanogr., 6, 879-893, 1976.

Kundu, P. K., S.-Y. Chao, and J. P. McCreary, Transient coastal currents and inertio-gravity waves, Deep Sea Res., 30, 1059-1082, 1983.

La Fond, E. C., Internal waves, I, in The Sea, vol. I., edited by M. N. Hill, pp. 731-751, Interscience, New York, 1962.

Maxworthy, T., A note on the internal solitary waves produced by tidal flow over a three-dimensional ridge, J. Geophys. Res., 84, 338-346, 1979.

McEwan, A. D., Internal mixing in stratified fluids, J. Fluid Mech., 128, 59-80, 1983.

Millot, C., and M. Crépon, Inertial oscillations on the continental shelf of the Gulf of Lions - Observations and theory, J. Phys. Oceanogr., 11, 639-657, 1981.

Nagashima, H., Reflection and breaking of internal waves on a sloping beach, J. Oceanogr. Soc. Japan, 27, 1-6, 1971.

Petrie, B., $M_2$ surface and internal tides on the Scotian shelf and slope, J. Mar. Res., 33, 303-323, 1975.

Pollard, R. T., On the generation by winds of inertial waves in the ocean, Deep Sea Res., 17, 795-812, 1970.

Pollard, R. T., Properties of near-surface inertial oscillations, J. Phys. Oceanogr., 10, 387-398, 1980.

Pollard, R. T., and R. C. Millard, Comparison between observed and simulated wind-generated inertial oscillations, Deep Sea Res., 17, 813-821, 1970.

Prinsenberg, S. J., W. L. Wilmot, and M. Rattray, Generation and dissipation of coastal internal tides, Deep Sea Res., 16, suppl., 179-195, 1974.

Rattray, M., On the coastal generation of internal tides, Tellus, 12, 54-62, 1960.

Rattray, M., J. Dworski, and P. Kovala, Generation of long internal waves at the continental slope, Deep Sea Res., 16, 179-195, 1969.

Sawyer, C., Tidal phase of internal wave generation, J. Geophys. Res., 88, 2642-2648, 1983.

Sawyer, C., and J. Apel, Satellite images of ocean internal waves, Rep. NOAA S/T 2401, U.S. Dept. of Commerce, Washington, D.C., 1976.

Schott, F., Spatial structure of inertial-period motions in a two-layered sea, based on observations, J. Mar. Res., 29, 85-102, 1971a.

Schott, F., On horizontal coherence and internal wave propagation in the North Sea, Deep Sea Res., 18, 291-307, 1971b.

Schott, F., On the energetics of baroclinic tides in the North Atlantic, Ann. Geophys., 33, 41-62, 1977.

Summers, H. J., and K. O. Emery, Internal waves of tidal period off Southern California, J. Geophys. Res., 68, 827-839, 1963.

Tang, C. L., Inertial waves in the Gulf of St. Lawrence: A study of geostrophic adjustment, Atmos. Ocean, 17, 135-156, 1979.

Thomson, R. E., and W. S. Huggett, Wind-driven inertial oscillations in Queen Charlotte Sound and Hecate Strait, May-Sept., 1977, Pac. Mar. Sci. Rep. 81-20, 90 pp., Inst. of Ocean Sci., Sidney, B.C., 1981.

Torgrimson, G. M., and B. L. Hickey, Barotropic and baroclinic tides over the continental slope and shelf off Oregon, J. Phys. Oceanogr., 9, 945-961, 1979.

Wang, D-P., and C. N. K. Mooers, Evidence for interior dissipation and mixing during a coastal upwelling event off Oregon, J. Mar. Res., 35, 697-713, 1977.

Winant, C. D., Coastal current observations, Rev. Geophys., 17, 89-98, 1979.

Winant, C. D., and A. W. Bratkovich, Temperature and currents on the Southern California shelf: A description of the variability, J. Phys. Oceanogr., 11, 71-86, 1981.

Wolanski, E., and G. L. Pickard, Upwelling by internal tides and Kelvin waves at the continental shelf break on the Great Barrier Reef, Aust. J. Mar. Freshwater Res., 34, 65-80, 1983.

Wunsch, C., Internal tides in the ocean, Rev. Geophys., 13, 167-182, 1975.

Wunsch, C., and R. Hendry, Array measurements of the bottom boundary layer and the inertial wave field on the continental slope, Geophys. Fluid Dyn., 4, 101-145, 1972.

# SHELF BREAK CIRCULATION PROCESSES

John A. Johnson and Nicole Rockliff[1]

School of Mathematics and Physics, University of East Anglia, Norwich NR4 7TJ, England

**Abstract.** This paper concentrates on special features in the current, density, and nutrient distributions that occur in the neighborhood of the shelf break. Observations of shelf edge effects from the continental margins around the world are described. Analytical and numerical models that include shelf break topography are reviewed and compared with observations.

## 1. Introduction

Oceanography observations are made around the edges of many of the world's oceans and seas, with particularly large amounts of data being collected from the interesting areas associated with western boundary currents or with coastal upwelling along the eastern boundaries of oceans. In most of these regions there are continental shelves which enhance coastal currents and trap internal waves. Often there is quite a sharp transition in bottom relief from the gently sloping continental shelf to the steeper slope region, with the change in gradient taking place over a relatively short distance. This transition zone is often referred to as the shelf break.

The sharpness of the shelf break varies with distance along the shelf in a given region but often varies considerably more between regions. This is illustrated in Figure 1, which shows typical shelf-slope topographies for sections off northwest Africa, Oregon, and Peru (presented by Smith [1981] in a paper that compares the characteristics of coastal upwelling in the three areas). Off northwest Africa, there is generally a broad shelf with a sharp edge, whereas off Oregon the transition from shelf to slope is gradual, and the shelf break is a fairly indistinct feature of the bottom topography. The Peruvian shelf is often narrow but with a fairly distinct shelf break.

The character of the local topography clearly will affect the features of shelf break dynamics, both observed and modeled, that are discussed in

this paper. These include secondary upwelling and shear in the longshore current over the shelf break, two- or three-dimensional balances in the onshore transport near the shelf edge, poleward undercurrents that are frequently associated with the slope just below the shelf break in upwelling regions, and the formation of fronts in the neighborhood of the edge of the shelf. Such fronts often form on both eastern and western continental shelves and separate the shelf waters from the deep ocean; markedly different properties of temperature, salinity, and nutrient concentration are observed in the adjacent water masses. This role of the shelf break zone as a dividing region between two different regimes was commented on by Bowden [1981] during a Royal Society discussion meeting on shelf seas in 1981. He said that for the circulation on the shelf the deep ocean is involved as a driving force, the interaction taking place across the shelf break; its extent, however, is not often determined. From the oceanic side, the shelf break region can appear as a boundary layer of the deep-sea circulation. This interpretation is particularly applicable where western boundary currents flow along the continental slope just offshore of the shelf break, usually in association with a sharp front separating the coastal and boundary current water masses, as mentioned above.

Most observations of these phenomena have been made by conventional shipborne and moored instrumentation. Satellites provide an alternative means of observing some shelf break effects under certain conditions. For example, a clear image of the Celtic Sea front was obtained on June 19, 1979, from both TIROS N and NOAA 6 satellites, with the front showing up as a narrow ribbon of sharp temperature contrast near the shelf edge. Laboratory experiments have not been used extensively in ocean modeling, but a series of experiments by Whitehead [1981] include a model of the circulation in shallow seas with a shelf and shelf break. At moderate rotation rates a jet flows along the shelf break. At higher rates a distinct thermal front is formed at the shelf break, hindering sizable mass flux across the shelf edge; the front itself wanders around in association with turbulent eddies. These features closely

---

[1] Now at Department of Mathematics, Statistics and Computing, Plymouth Polytechnic, Plymouth PL4 8AA, England.

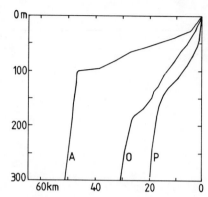

Fig. 1. Shelf and slope topography off northwest Africa (21°40'N), Oregon (45°N) and Peru (15°S) [from Smith, 1981].

resemble many of the observed results, which are discussed in detail in section 2. The presentation of these is on a geographical, regional basis, to deal with observations from many parts of the world.

Modeling of the ocean circulation to include the effects of the shelf break requires the incorporation of a number of features. Foremost, of course, is the bottom topography with shelf, slope, and deep ocean regions. In addition, the observed flows are often highly stratified and certainly usually very variable in time, with upwelling events caused by a few days of upwelling favorable winds. In the relatively shallow shelf water the effects of bottom friction are significant. Ideally, therefore, models ought to contain realistic topography, stratification, friction, and variability. Unfortunately, this combination produces a very difficult theoretical problem, and it is often necessary for one or more of these features to be simplified or omitted if theoretical progress is to be made.

Time scales for shelf break modeling offer interesting choices to the modeler because of the differences between the shelf and deep ocean regimes. The shelf flow, in general, is spun up on a relatively short time scale, typically less than a day for the longshore flow and about a week for the density field; these are both driven by the wind stress, which itself may be changing on a very short time scale of a few hours. The deep ocean, however, spins up relatively slowly, and changes forced by the wind may take weeks to months to have a lasting effect. This difference in time scales on either side of the break makes it likely that some frontal processes will occur in its neighborhood. For example, Hsueh et al. [1982] show that eddies are generated along the shelf edge when transport is forced across the shelf break by a wind-driven circulation on the shelf. A survey of theoretical models is given in section 3, with particular reference to time scales in sections 3.2 and 3.5 Analytical models are discussed in sections 3.1-3.5, and fully numerical models in sections 3.6 and 3.7.

In the descriptions that follow, the coordinate system is always such that x is positive eastward, y northward, and z vertically upward from the mean sea surface at z = 0. The sea surface may be displaced a small amount to $z = \eta(x, y, t)$, and the ocean bottom is at $z = -h(x, y)$. The corresponding velocities are u eastward, v northward, and w upward, and the reduced pressure is denoted by p. The equivalent temperature θ is defined by

$$\alpha\theta = \alpha(\text{temperature}) - \bar{\alpha}(\text{salinity})$$

where α is the coefficient of thermal expansion and $\bar{\alpha}$ is the fractional inverse in density per unit increase in salinity. The horizontal momentum equations used in most models can be derived from

$$u_t + (uu_x + vu_y + wu_z) - fv = -p_x + \upsilon_H(u_{xx} + u_{yy}) + \upsilon u_{zz}$$

$$v_t + (uv_x + vv_y + wv_z) + fu = \upsilon_H(v_{xx} + v_{yy}) + \upsilon v_{zz}$$

where $f = f_0 + \beta y$ is the usual Coriolis parameter for a β plane (f plane if β = 0) and $\upsilon_H$ and $\upsilon$ are horizontal and vertical eddy viscosities respectively. The hydrostatic balance is

$$p_z = g\alpha(\theta - \theta_0)$$

where g is the acceleration due to gravity and $\theta_0$ is a reference equivalent temperature. The continuity equation is

$$u_x + v_y + w_z = 0$$

and the equivalent heat equation may be written in the form

$$\theta_t + (u\theta_x + v\theta_y + w\theta_z) = \kappa_H(\theta_{xx} + \theta_{yy}) + \kappa\theta_{zz}$$

where $\kappa_H$ and $\kappa$ are horizontal and vertical eddy diffusivities.

In many applications to continental shelves and slopes, at either eastern or western ocean boundaries, the longshore length scale is much greater than the cross-shelf and vertical scales, and the longshore velocity is much larger than the onshore and vertical velocities. Thus

$$v \gg u, w \qquad \frac{\partial}{\partial y} \ll \frac{\partial}{\partial x}, \frac{\partial}{\partial z}$$

Furthermore, the nonlinear momentum terms are often small compared with the Coriolis acceleration terms (that is, a small Rossby number limit), and the lateral friction terms are usually less important than the vertical friction terms which bring in the bottom friction and surface wind stress. The advection of heat by the circulation is generally more important than the diffusion of heat. Therefore, for many models, the above equations may be simplified to

$$fv = p_x + \upsilon u_{zz}$$

$$v_t + fu = -p_y + \upsilon v_{zz}$$

$$p_z = g\alpha(\theta - \theta_0)$$

$$u_x + v_y + w_z = 0$$

$$\theta_t + u\theta_x + v\theta_y + w\theta_z = 0$$

The usual boundary conditions at the surface are

$$u_z \propto \tau^x \qquad v_z \propto \tau^y \qquad w = 0 \qquad \theta_z = Q$$

where $\tau^x$, $\tau^y$ are the eastward and northward components of surface wind stress and $Q$ is the surface heat flux. At the bottom and other impermeable boundaries the conditions are

$$u = v = w = 0 \qquad \underline{n} \cdot \nabla\theta = 0$$

where $\underline{n}$ is a vector normal to the boundary.

As each model discussed below incorporates its own particular modifications and approximations to the basic equations and boundary conditions, it is not feasible to list the appropriate forms each time: they may be found in the original papers. A brief description of the approximation in each paper (or set of papers) will, in general, be given.

Coastal-trapped waves, which are affected by the shelf-slope topography, are not discussed here, either in the observations or in the theoretical models, as they are the subject of a paper by Huthnance et al. [this volume]. Waves and currents near the continental shelf edge have also been described previously by Huthnance [1981]. A further recent survey which includes much on shelf circulation and has some discussion of shelf break effects is by Allen et al. [1983]; this forms part of the U.S. national report to the IUGG and describes the latest work on the physical oceanography of continental shelves in many parts of the American continent.

## 2. Observations

During recent years there have been numerous cruises to carry out observations in upwelling areas around the world, often as part of international programs. The most extensive observations have been made in the Pacific Ocean, off Peru in the south and Oregon in the north, and, in the Atlantic Ocean, off northwest Africa. More restricted local observational projects have been mounted around much of the ocean margins where coastal upwelling and hence high primary production have been expected.

Relatively few of these programs have made extensive observations of the region around the shelf edge, where it is believed that secondary

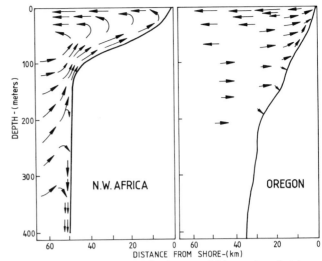

Fig. 2. Schematics of the upwelling circulation for northwest Africa (21°40'N) and Oregon (45°15'N) [from Huyer, 1976].

upwelling may occur. Many observational experiments use the shelf edge as a convenient offshore boundary for coastal oceanography measurements. Therefore detailed observations with sections across the shelf edge with stations at and close to either side of the break are fairly rare. Those that have been detected in the literature are included in the following subsections.

A comparison of upwelling events off northwest Africa and Oregon, made by Huyer [1976], shows some of the diversity of local characteristics and observed phenomena. Figure 2 illustrates the conditions encountered in JOINT I and CUE 2 experiments. Off northwest Africa at 21°40'N there is a broad shelf with a sharp shelf break and fairly weak stratification. Off Oregon, however, the transition from shelf to slope is gradual, and the stratification is strong. The onshore and upward currents near the shelf break are dominant features of the northwest African observations but are virtually absent over the weak break off Oregon. Huyer [1976] also points out that the undercurrent generally remains over the slope off northwest Africa but penetrates over the shelf off Oregon. For northwest Africa the coldest water and the greatest changes in density in an upwelling event occur just inshore of the shelf break, whereas for the Oregon shelf these effects are confined to the nearshore region. Further details of both regions are presented in sections 2.1 and 2.3.

Some properties that feature in many observations from the neighborhood of shelf breaks around the world are the existence of a front between the shelf water and the deep ocean waters, the three dimensionality of the circulation near the edge of the shelf with a local absence of onshore-offshore balance in the mass transport, and the presence of a slope current just seaward of the break. These properties are described in the following subsections, in which the observations have been organized on a geographic basis. The first four subsections deal with the well-known upwelling regions along the eastern boundaries of the Pacific and Atlantic oceans. The fifth deals with the shelf edge just inshore of the Gulf Stream, and the final section describes a few observations from regions as far apart as the Celtic Sea and the Great Barrier Reef.

## 2.1. Eastern North Pacific

One of the most intensively studied upwelling areas is the continental margin off California, Oregon, Washington, and Vancouver Island. This is not, however, a very productive region for interesting observations over the shelf break, probably because the gradients of the bottom topography change slowly with distance from the coast and there is not a very sharp break between the continental shelf and the continental slope.

Starting with the northernmost part of the eastern boundary of the North Pacific, Freeland and Denman [1983] report on an annual upwelling event that occurs near Vancouver Island. In spring, anomalously high $\sigma_t$ and low-oxygen levels are observed along the shelf. The seaward side of the anomalies is bounded by a sharp frontlike region which roughly follows the edge of the continental shelf. In winter, prior to the event, the shelf edge currents are northbound, but during the spring transition this current reverses. At all seasons the currents are strongly polarized parallel to the local bathymetry. It is believed that this upwelling event is associated with the Juan de Fuca canyon just south of Vancouver Island and that the anomalous water is not upwelled across the shelf break but comes up the canyon from below the break.

Further observations of the currents along the Pacific coast of Canada are described by Freeland et al. [1984]. Using current meter and conductivity/temperature/depth (CTD) data from different parts of the British Columbia coast, they have produced cross sections of alongshore current extending from the coast to over 1000 m depth. In summer a southward flowing current is observed over the shelf break with its maximum just seaward of the shelf edge. The jet appears to be a response to the local wind system, and its location to be related to bottom topography. There is also a narrow coastal current that may be driven by a local runoff.

An upwelling event off Vancouver Island between July 21 and August 5, 1980, has been described by Ikeda and Emery [1984] from satellite infrared images. Before the intensification of the upwelling favorable wind from the northwest there was a cold band of water just over the continental shelf break. Nearly simultaneously with the increase in the wind speed, colder sea surface temperatures began to appear near the coast. This cold band broadened, and the offshore boundary propagated

seaward out to the shelf break at a speed of about 10 km/d. After the wind diminished at the end of the event, warmer water appeared again at the coast, with the cold band remaining over the break. The authors say that this cold band might indicate shelf break upwelling but could be cooler water advected from further north.

Measurements collected off Washington in the autumn of 1971 by Cannon et al. [1975] show a regular northward flow over the continental slope below the edge of the shelf. Although the observations spanned the shelf break, there were no significant horizontal gradients at depths shallower than the shelf break. This current may be a continuation of the California undercurrent along the slope. Deepening of the density surfaces at the continental slope indicates that the current may be intensifying below the shelf break.

Observations from the Oregon shelf do not contain many references to interesting shelf break effects. Huyer [1976] compares the upwelling events off Oregon with those off northwest Africa and says that on the Oregon shelf most of the interesting hydrographic effects occur near the coast. Johnson and Johnson [1979] say that a weak double-cell circulation sometimes appears over the shelf break off Oregon. In summer the deep northward flow is strongest at the outer edge of the shelf, as noted by Huyer et al. [1975], but there are not significant shelf edge effects in other seasons.

The California current system is reviewed by Hickey [1979], who provides a very extensive summary of observational data but only briefly mentions shelf edge phenomena. One noticeable feature is a poleward undercurrent with maximum velocities just offshore of the California shelf break. Current meter data for moorings at various depths over the shelf and the slope off southern Washington are presented as time series for July to September 1972. The mean alongshore currents during this period are illustrated on a cross section and show a northward flow with maximum just offshore of the shelf break and extending onto the shelf. The mean flow near the surface is southward. The poleward undercurrent is also observed near the shelf edge off Oregon by Huyer and Smith [1974]. In summer the coastal equatorward jet often extends out beyond the shelf break, as the shelf is narrow, but it is confined to the surface layers.

Some winter measurements off Washington are also described by Hickey [1979] with current meters above midshelf, shelf edge, and slope with evidence of strong currents above the shelf edge and midshelf. A warm core eddy has been observed by Huyer et al. [1984] off Oregon in winter 1978. The eddy penetrates down to more than 800 m over the continental slope but does not reach the bottom. The presence of this eddy for at least 2 months dominates the annual mean alongshore flow.

At the southernmost part of this region in the north Pacific is the coast of Baja California. Observations reported by Barton and Argote [1980] show a well-defined undercurrent flowing north along the continental slope. During the observations the poleward undercurrent penetrated at times into the outer and midshelf regions. The shelf here is narrow with a more definite shelf break than occurs further north, but no particularly noteworthy edge effects are reported. Apart from the short time scale sea breeze cycle, the winds are almost always favorable to upwelling. This lack of long-term variability in the wind leads to little expectation of shelf break upwelling activity (see discussion of the work by Johnson and Nurser [1983] in section 3.2).

In summary, observations of the eastern coast of the north Pacific to date forecast only one interesting shelf edge feature. That is the northward flowing undercurrent close to the continental slope just below the shelf break. This current, during strong upwelling, occasionally spreads onto the shelf, as described by Huyer and Smith [1974].

## 2.2. Eastern South Pacific

The Peru upwelling region has attracted much attention, with particular reference to the changes occurring during El Niño. Brink et al. [1980], in forecasting observations from the 1977 JOINT II project, say that the upwelling defies a simple description, especially near the shelf break, due to its strongly three-dimensional character. The persistent imbalance of the cross-shelf volume flux is a curious aspect of the system and may be due to nonuniformities in bottom topography and longshore variations of wind stress, as suggested by the numerical model of Preller and O'Brien [1980].

From the same series of JOINT II measurements, Brockmann et al. [1980] find that the mean geostrophic velocity profiles at 15°S show a maximum poleward flow between 50 and 150 m depth and that the undercurrent is confined to the slope region within 50 km of the shelf break, that is, within a baroclinic radius of deformation. These effects are shown in Figure 3, where a significant shear in the longshore flow above the shelf break is also apparent.

In a review of the physical environment of the Peruvian upwelling system, Brink et al. [1983] describe further many observations made during JOINT II. They reiterate that a two-dimensional mass balance does not hold anywhere and that from the shelf break seaward, even the qualitative resemblance to two dimensionality and steady upwelling breaks down. The onshore compensation flow is restricted to the region inshore of the shelf edge. Coastal-trapped waves are an important feature in this region. The maximum variance associated with these waves appears at about 200 m near the shelf break.

To summarize the shelf edge effects on the Peruvian shelf, the most common observation is the lack of two dimensionality in the mass balance and therefore the obvious need for a fully three-di-

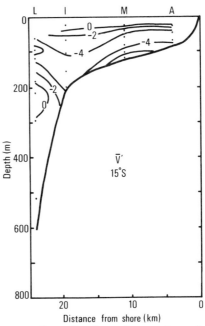

Fig. 3. The distribution of mean longshore flow from current meter arrays at 15°S. The negative values (centimeters per second) indicate poleward flow [from Brockmann, et al. 1980].

mensional model. A slope countercurrent is again an important feature.

Observations in the Humboldt current off Chile have been made by Johnson et al. [1980]. They show the slope countercurrent continuing in this more southerly region. The flow is again very three dimensional but with some evidence of double-cell cross-shelf circulation, strong in midshelf but rather weak at the shelf break. There is some evidence that water upwells across the shelf break from the slope countercurrent onto the shelf.

## 2.3. The Northwest African Shelf

By far the most observations showing interesting shelf break phenomena originate in the coastal waters off Mauritania and Senegal, in northwest Africa. During a 1969 cruise, measurements were made by Hughes and Barton [1974] which showed the isopycnals well aligned with the shelf edge from Cabo Bojardor to Cap Blanc. The distributions of salt and heat on the $\sigma_t = 26.8$ surface indicate a fairly narrow stream of cooler, less saline water flowing toward the north at a depth of 200-300 m parallel and close to the shelf edge. The vertical extent of the influence of this flow is limited, as there are no similar indications of a poleward movement in the distributions on the density surfaces $\sigma_t = 26.7$ and $\sigma_t = 26.9$. Also using data obtained in 1969, Jones [1972] describes sections of salinity, temperature, nitrate, and silicate,

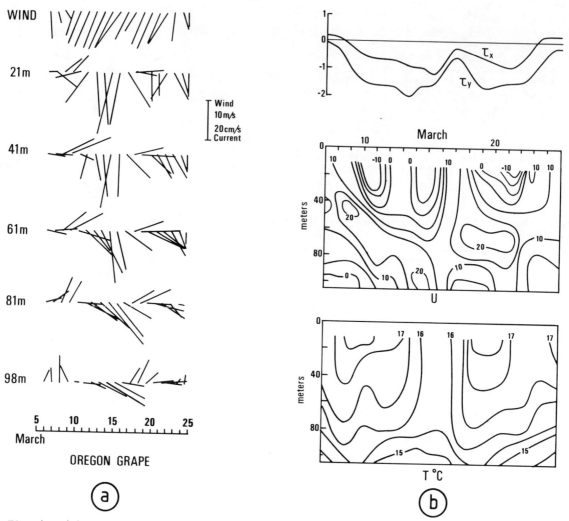

Fig. 4. (a) Current vectors at Oregon Grape (a shelf edge mooring) and wind vectors on the shelf. (b) wind stress (dynes per square centimeter), onshore velocity u (centimeters per second), and temperature (degrees Celsius) at Oregon Grape [from Mittelstaedt et al., 1975].

which show distinct upward protrusions of the isohalines over the edge of the shelf. There is some evidence of a front just inshore of the shelf break; the $\sigma_t$ measurements, however, do not show this particularly, although they do show coastal upwelling.

A number of observations near the shelf break were made by Mittelstaedt et al. [1975] as part of the JOINT I project. They observed that the most intense ascending motions, of $0(10^{-2}$ cm s$^{-1})$, occur between the shelf edge at 100 m depth and the top of the poleward undercurrent at 150 m. This is associated with the maximum onshore flow occurring close to the shelf break and further shoreward. The flow is not two dimensional, and the boundary zone between the offshore countercurrent and the wind-driven coastal regime stays near the shelf break. Ascending water from depths

below the break is first absorbed by the accelerating onshore flow and then upwells into the surface layers inshore of the shelf edge. Figure 4a shows the longshore velocity v compared with the wind stress, both measured at the shelf break station called Oregon Grape. The velocity field at the shelf break is very variable in time compared with the fairly uniform wind, so the flow along the shelf edge is clearly not well correlated with the wind.

Figure 4b shows the onshore flow u and temperature θ compared to the wind at Oregon Grape station over a period of 3 weeks. Note that a two-dimensional balance in onshore-offshore flow occurs only at certain times, usually after a long spell of upwelling favorable winds when the coastal wind effect has penetrated to the shelf edge. Notice also the colder water at the bottom that

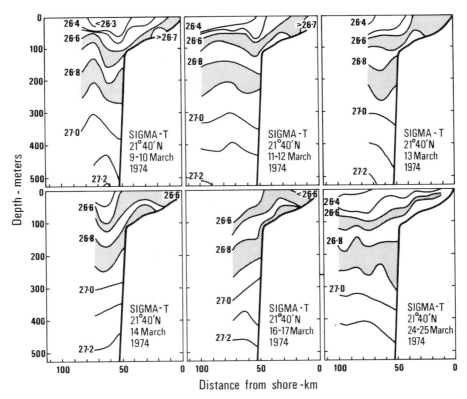

Fig. 5. Time series of sigma-t sections along 21°40'N showing the development during a major upwelling event and subsequent relaxation [from Barton et al., 1977].

has come from deeper offshore. Temporal variations in the flow near Cabo Corveiro are described by Barton et al. [1977], who point out that the response to the wind is essentially confined to the shelf. Upwelled water appears initially over the inner shelf, but after continuing upwelling favorable winds the center of upwelling migrates to the shelf break. Despite the persistence of upwelling favorable winds, the cold water migrates no further seaward than the shelf edge, where it remains until the winds weaken. This is illustrated in Figure 5 for the period March 9 to March 25, 1974, with the winds essentially as given in Figure 4 at the shelf break. Jones and Folkard [1970] also observed the coolest, freshest, highest-nitrate, lowest-oxygen surface water situated over the shelf break in early March 1966 near Cap Blanc. Meincke et al. [1975] found a band of cold, low-salinity water along the shelf break at the latitude of, and south of, Cap Blanc in February/March 1972.

As part of the JOINT I experiment, Johnson [1976] used a profiling current meter in the section at 21°40'N with measurements on both sides of the shelf break. Combining hydrographic observations with current profiles, the following deductions about the flow are made. There is upwelling into the surface layer between midshelf and the shelf break. There is a seaward flow, near surface above the shelf break, which descends within the weak frontal region, creating a temperature and salinity minimum there. Longshore advection is thought to be important in maintaining the density distribution on the shelf, and the shallow depth of the shelf break seems effectively to prevent deeper water from intruding onto the shelf. Hagen [1981] also noted that the wind-generated offshore current induced downwelling at the main density front along the shelf edge. This front with isopycnals upwarping to the surface is embedded in the strong geostrophic surface current parallel to the coast.

The Auftrieb 1977 program investigated the shelf off Mauritania and is described by Mittelstaedt and Hamann [1981]. The shelf break mocrings appeared to be deployed in a transition zone between the southward shelf flow and the northward offshore currents. As the zone shifted slightly, abrupt changes occurred in the current measurements. Convergence in the currents toward the transition zone favored the formation of a front and the downwelling of water from both inshore and offshore. The front with downwelling appears as a depression in both isopycnals and isotherms. This frontal downwelling combined with coastal upwelling provides the possibility of water and plankton recirculation. The pronounced horizontal gradients in the surface layer, which intensify at times into fronts in the vicinity of the shelf break, characterize the transition between the

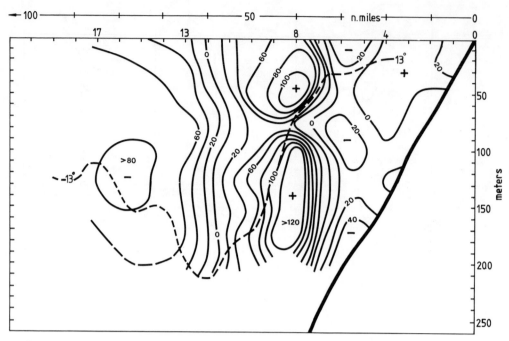

Fig. 6. Longshore velocity (centimeters per second) off Cape Town, South Africa, showing frontal jet near edge of shelf [from Bang and Andrews, 1974].

cool coastal waters and the relatively warm mixed oceanic water offshore.

A summary of the northwest African upwelling region has been given in a review by Mittelstaedt [1983]. He states that the frontal zone is a highly variable structure that occurs mostly in the vicinity of the shelf edge, with the countercurrent just offshore of the shelf break. Sometimes the northward flow encroaches from offshore onto the shelf. The shoreward advance of this flow starts at subsurface depths along the shelf break.

Occasionally, the freshest upwelled water has been detected away from the coasts over the mid-shelf or even at the shelf break. The most natural explanation of this situation appears to be alongshore advection of different water masses. However, this should not preclude the possibility that cool subsurface water sometimes upwells to the surface at the shelf break and over mid-shelf. The convergence of streamlines normal to the slope implies relatively strong vertical motions right at the shelf break, just below the compensation layer. However, most of the water ascending from below the shelf edge becomes trapped in the onshore flow within the compensation layer and upwells into the surface layer inshore of the break.

## 2.4. The Southwest African Shelf

This region is characterized by the cool waters of the Benguela current flowing equatorward along the southern Atlantic coast of the African conti-

nent and has long been recognized as a significant upwelling area. It attracted the attention of Defant [1936], who both compiled a systematic picture of the currents and provided one of the earliest theoretical models which gave some qualitative explanation for the observed circulation. Observations of the Benguela region have been somewhat infrequent, and the data obtained rather sparse; fortunately, however, the observations in this area, particularly by Hart and Currie [1960], Bang [1971], and Bang and Andrews [1974], have extended over the shelf break zone, which is fairly well defined here.

The location of a region of upwelling or disturbance some 60-80 miles (96-128 km) offshore was noticed by Bang [1971]. The ridging of the isotherms implied horizontal divergence at the surface and a compensating upward movement from below to satisfy continuity. This surface divergence was a characteristic of the flow above the shelf edge between the Orange River and about $13 \frac{1}{2}$S. Hart and Currie [1960] forecast evidence of a similar feature continuing northwestward. There was also clear evidence of sporadic shelf edge divergence west of Cape Town. This divergence associated with the discontinuity in continental slope separates the weather-dominated Benguela system from the climate-dominated southeast Atlantic deep-sea circulation. The shelf edge activity took place in the absence of coastal upwelling and was independent of local wind forcing.

In a later paper, Bang and Andrews [1974] investigated a strongly baroclinic frontal zone near the shelf break containing an intense equatorward jet of maximum velocity 1.2 m/s. Inshore upwelling was not taking place at the time. The jet illustrated in Figure 6 appears to be geostrophically adjusted, and the energy required to maintain it small, so during spells of prolonged upwelling favorable wind stress, the potential energy accumulates at a rate faster than the frontal jet can utilize it. As a result the front moves offshore until it reaches the shelf edge. The jet can then expand downward and its energy-containing capability increases so that an equilibrium situation may become established somewhere near the shelf edge.

To summarize, observations are rather sparse over the southwest African shelf, but the inference from isotherm patterns is that shelf break upwelling may well exist if associated with a frontal zone and a frontal jet.

## 2.5. The Western North Atlantic

Along the eastern coast of the United States and Canada there have been many hydrographic surveys across the edge of the continental shelf. In the south these have been in connection with the Gulf Stream, which follows the continental slope. Many observations suggest that the shelf break is associated with the position of the front separating shelf waters from deep ocean waters.

Leming and Mooers [1981] interpreted the upward bowing of the isotherms over the shelf break off Cape Canaveral during onshore movement as bottom-trapped upwelling at the shelf break. The winds during these observations were always favorable to upwelling. From Cape Canaveral to Cape Hatteras in summer, Blanton et al. [1981] observed wavelike perturbations along the edge of the Gulf Stream which induced upwelling of colder, less saline, North Atlantic central waters at the edge of the shelf. The upwelled water intruded along the bottom and displaced large amounts of shelf water. Figure 7 shows upward bowing of the isotherms over the shelf break. The upwelling occurs more frequently over diverging bathymetry downstream of capes.

Near Cape Hatteras on the North Carolina shelf, Blanton [1971] reports that the divergence at the shelf break created by the frequent southwesterly winds often causes upwelling of Gulf Stream waters onto the shelf and that changes at the shelf break during a 24-hour period can be substantial. Accelerations of the Gulf Stream along the continental slope may cause similar intrusions. During summer conditions when the shelf water is stratified, intrusions of Gulf Stream waters occur along the bottom and involve water from the continental slope below the break, thereby lowering the temperature of the shelf bottom water.

Further north over the eastern Georges Bank, where the Gulf Stream has turned away from the coast, the Gulf Stream front usually intersects the surface within 50 km offshore of the break in the southwest and 150 km in the northeast. Beneath the surface it slopes downward and onshore,

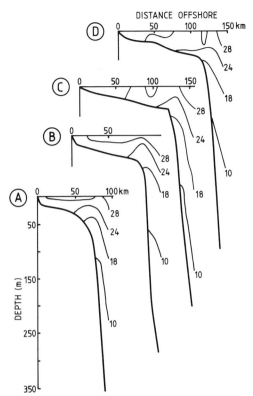

Fig. 7. Temperature transects on the continental shelf between New Smyrna Beach, Florida, and Savannah, Georgia. (a) New Smyrna Beach, (b) St Augustine, (c) Brunswick, and (d) Savannah (adapted from Blanton et al., [1981]).

generally intersecting the bottom just shoreward of the shelf break and forming a bottom front there, as described by Halliwell and Mooers [1979]. Near the shelf break, anticyclonic eddies form north of the Gulf Stream and force near-surface shelf water offshore at their northeast edge, thereby producing large-amplitude seaward perturbations of the shelf-slope front. Shoreward perturbations of the bottom front can be produced by eddies in their southwest quadrant. However, shoreward movement of the surface front was produced by only a few of the eddies examined by Halliwell and Mooers [1979].

Between Cape Charles and the Great South Channel, the shelf-slope front was bound to the shelf break except when eddies drove shelf water offshore. Along the eastern end of Georges Bank the front was not tightly bound to the shelf break and was frequently displaced offshore by up to 100 km for several weeks.

Over the Nova Scotia shelf, upwelling occurs at the shelf break from depths of 400 m with peak vertical velocities of 2 mm/s. Petrie [1983] shows that the shelf break circulation responds coherently to wind forcing (very different from eastern ocean boundary situations). In autumn, upwelling and strong onshore flow are significantly correlated with current pulses following the changes in temperature at the break. Temperature at the shelf break starts to fall immediately as the wind increases, but there is a delay of 18 hours for a corresponding drop at the shelf mooring, corresponding to an onshore velocity of 0.34 m/s, a strong onshore current. Vertical shear develops for both components of current at the shelf break, with the current decreasing with depth. There are strong longshore currents with large vertical shears confined to the shelf and shelf break regions, with intense upwelling observed at the shelf break.

## 2.6. Other Upwelling Areas

Some interesting observations of upwelling over the edge of the Celtic Sea shelf have been made using satellites. The most well known example is the June 19, 1979, infrared images by NOAA 6 and TIROS N shown by Pingree and Mardell [1981] and Dickson et al. [1980], respectively. A persistent but localized band of upwelling follows the contours of the European continental slope from Porcupine sea bight to the Bay of Biscay as indicated in Figure 8. The narrow ribbon of cool water is aligned almost exactly with the top of the continental slope. In general terms, the cooling increases in extent and intensity northwestward along the slope. A similar band appears on the May 14, 1979, TIROS N image.

Pingree and Mardell [1981] also describe some observations made in 1972-1973 across the Celtic Sea shelf break. They found that the surface waters over the slope were colder than the adjacent waters of the Celtic Sea and the Atlantic Ocean. Transects across the shelf break in the summers of 1976 and 1980 show a broad band of cooler water (60 km wide) centered over the slope region. Isolated high-nitrate patches occur, and increased chlorophyll concentrations are associated with the generally cooler water: this suggests some upwelling of nutrients. The CONSLEX experiment in 1982 along and across the shelf break zone northwest of Scotland should give more information about shelf break effects in higher latitudes, but data from the experiment are not yet available.

In the northwest Pacific Ocean there are wide shelves in the Gulf of Alaska and in the Bering Sea. The former region has received attention from Lagerloef et al. [1981], who describe a 3-year period of observations over the northeast Gulf of Alaska shelf break. They found that the mean flow was along the shelf at all depths, with negligible vertical shear. The variance in the current was high in both along-shelf and cross-shelf components, but the correlation with variations in the wind was low.

Analysis of data from the Bering Sea shelf has been made by Kinder and Coachman [1978] to describe the haline front that lies over the continental slope in the eastern Bering Sea. The inner band of the Bering slope current flows northwestward parallel to the shelf break. The shelf re-

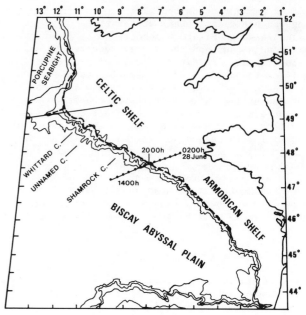

Fig. 8. Location chart showing the relation between the axis of the cold water belt (dashed line) and the 250-, 500-, and 1000-fathom (457-, 914-, and 1828-m) slope contours (adapted from Dickson et al. [1980]).

gion is dominated by tidal motion. The salinity gradient appears to be higher near the shelf edge than further offshore. The front does not show up in temperature sections. They conclude that the major requirements for the formation of similar fronts are a broad and shallow continental shelf, significant runoff at the coast, a shelf regime dominated by high frequencies, and a strong effect of salinity on density. This front forms a boundary between the tidally dominated shelf regime and the geostrophic deep-water regime. Further work on fronts in this region has been carried out by Shumacher and Kinder [1983]. They note that the shelf and oceanic domains are separated by the 50-km-wide shelf break front. A weak west wind induces a boundary current parallel to the slope, and eddies frequently occur seaward of the shelf break. The shelf region is broken up by various fronts which occur above rapid changes in shelf gradient, as if there are three separate shelf break fronts.

At the opposite end of the Pacific Ocean on the Queensland shelf off northeast Australia, there is localized upwelling of enriched bottom water from the shelf break onto the reefs. This upwelling reported by Andrews [1983] is important for the reefal ecosystems.

This section on observations has described features of the circulation near the shelf break around the world. Clearly, in some upwelling areas the observations in the neighborhood of the edge of the shelf are rather sparse. In the next section a variety of theories are discussed and compared with some of the above descriptions. Some comments about the need for further observations are made in section 4.

## 3. Theory

Theoretical work on flows over the shelf began in the early 1970s and has continued to the present day, with more physical properties being included as the models become more sophisticated. From the observations described in section 2 it is clear that a good model of shelf break upwelling has to include certain basic properties. Three of the most important are (1) obviously the bottom topography, which should include a shelf with a fairly rapid transition of gradient at the shelf edge to the steeper continental slope, (2) variability of the flow being driven by wind stresses that change with time to represent upwelling events, and (3) stratification of the flow to model the characteristic upward bowing of the isotherms and to cope with the density fronts that often form in the neighborhood of the shelf edge.

One point of difference that appears in the models described in this section is the choice of f plane or β plane dynamics. Some authors claim that the f plane is sufficient, as the topographic effects dominate changes in f/h. This is often true for very local situations in which the appropriate longshore horizontal length scale is small by the choice of restricted wind forcing or geographical extent, for example, a single bay. Other authors use a β plane; this is particularly necessary when the shelf circulation is matched to deep water flows offshore which vary on a long north-south scale.

Time scales are also a source of variety in the models. Some analytical models are steady and try to represent the situation attained asymptotically long after spin-up has ceased. Some models change only slowly with time, whereas others use different time scales for the shelf and deep water flows or different time scales for the barotropic and baroclinic components. In general, the baroclinic and deep water flows are slower to spin up than the barotropic and shelf water flows.

Numerical models have been used widely to treat this complex flow. They range from single-layer models through two- and three-layer representations to continuous stratification. Some are fully numerical in the sense that they solve the basic primitive equations using finite difference schemes. In other models, analytic techniques are first used to reduce the equations to a simpler set, or eigenfunction expansions are introduced to represent some of the spatial variations; numerical solution of the simplified problem then follows. For these numerical models, the sharpness of the shelf edge and the steep gradient of the slope often pose several numerical problems.

One reason for the wide variety of models is the difficult nonlinear set of equations that arise when topography, stratification, and variability are all included. Many shelf flow models include only one or two of these basic features. As this paper concerns the shelf break circulation, all models described here include the bottom topography but may or may not include stratification and variability.

A review of models of coastal currents on continental shelves has been presented by Allen [1980]; although there is no discussion of shelf edge effects, it is useful for a general overview of some of the theories that have been extended to include a shelf break. Huthnance [1981] has surveyed theories of waves and currents near the continental shelf edge. He points out that it is often convenient to model ocean or shelf separately, with shelf edge boundary conditions representing the transmission, trapping, or reflection of various motions. This approach is apparent in some of the models discussed below.

The summary of the various models of shelf break dynamics presented here has been subdivided for ease of description, using three methods of classification: steady ($v_t = \theta_t = 0$) versus unsteady ($v_t$ or $\theta_t \neq 0$); homogeneous ($s = 0$) versus stratified ($s \neq 0$); analytical versus numerical. Another natural division between models is linear and nonlinear. This classification, however, is split among the subsections according to the other properties of the models.

### 3.1. Steady Barotropic Models

This set of models omits two of the basic requirements of a good model, namely, stratification and variation with time. However, their authors reasonably claim that the simple model is a starting point from which more complicated models follow and that it often shows up some of the basic physics involved better than too complicated models which have to be solved numerically.

In the first of a series of papers, Hill and Johnson [1974] investigated the effect on the circulation of a sharp discontinuity in gradient at the edge of a flat continental shelf in a β plane wind-driven ocean model. The method of solution involves the construction of shear layers above the shelf break, to enable the deep water circulation to be matched to that on the continental shelf. The model is three dimensional, linear, and driven by a steady wind stress. There is no significant northward transport in the shear layer, its main purpose being to smooth out the discontinuity in the longshore velocity v between the deep water and the shelf circulation. Associated with this shear in v is recirculatory secondary upwelling above the shelf break, with an upward jet between two compensatory downward flows, as suggested by Bang [1971] in his observations of the Benguela current. Upwelling occurs at the shelf break when the u velocity is onshore. This theory was modified by Johnson and Killworth [1975] to show that the vertical circulation is strongest near the bottom of the shear

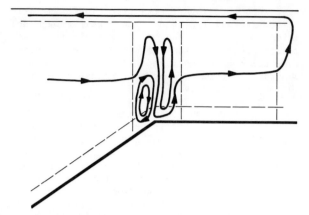

Fig. 9. The transverse flow over the shelf break with onshore flow in the interior [from Johnson and Killworth, 1975].

layer at the shelf break (as shown in Figure 9) and that a bottom current forms along the shelf break as a consequence of the discontinuity in Ekman transport there. The arguments used to predict the bottom current are independent of the concept of eddy viscosity used in the theory.

A model that introduces a continuous gradient shelf break is described by Lill [1979]; the bathymetry consists of separate shelf and slope sections smoothly connected at the shelf break. This two-dimensional model on an f plane, with no dependence of the velocity field on the longshore coordinate y, consists of a surface Ekman layer above a deep return layer. It is assumed that the offshore velocity gradients become small at large distances away from the coast and that an onshore-offshore balance exists at each latitude. The model is driven by prescribed surface wind stress and allows a longshore pressure gradient to exist. As shown in Figure 10, the onshore flow forms a jet close to the bottom in the vicinity of the shelf break and over the slope, with a small circulation cell near the break. The maximum upwelling occurs at the coast, but there is a

secondary local maximum in w just seaward of the shelf edge and close to the bottom. The cross-shelf streamlines in Figure 10 may be compared with the observations off northwest Africa shown in Figure 2.

Returning to the sharp shelf break with a discontinuity in gradient, Johnson and Manja [1979] revised the theory of Hill and Johnson [1974] to allow for a shelf that is shallow compared with the deep ocean and therefore has stronger coastal currents over the shelf. In a very mathematical paper they demonstrated how the shallow shelf circulation can be matched to that in deep water by a complex shear layer over the shelf break. There is no net vertical transport in the shear layer. In an example driven by a steady longshore wind stress $\tau^y$, the theory predicts that the longshore jet formed above the break has northward transport in dimensionless variables proportional to

$$\alpha \tau^y f^{1/2} / (\beta^2 d^3)$$

where $\alpha$ is the change in slope at the shelf break and d is the depth at the break. If $\tau^y < 0$, there is coastal upwelling, and the transport at the shelf break is southward. Its magnitude increases as the strength of the wind increases, as the change in gradient becomes sharper, and as the depth of the shelf decreases. Based on this theory is Figure 11, from Johnson and Manja [1982]; this illustrates the cross-shelf flow and secondary upwelling associated with the longshore jet above the shelf break. The large velocities near the bottom are a notable feature, which may be compared with some of the illustrations of observations in the northwest African upwelling area, in the work by Mittelstaedt et al. [1975] (see Figure 4).

A recent paper containing shelf break topography is by Thompson and Veronis [1983], which presents a model of circulation in the Indian Ocean, around the shelves of Western Australia, using a homogeneous linear steady β plane model. The equations of motion are reduced to a single

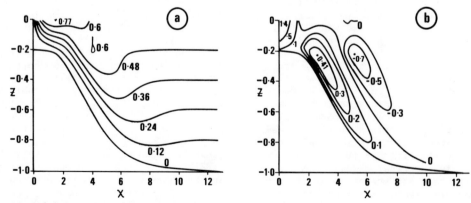

Fig. 10. (a) Dimensionless streamlines and (b) contours of dimensionless vertical velocity, in the return layer below the Ekman layer [from Lill, 1979].

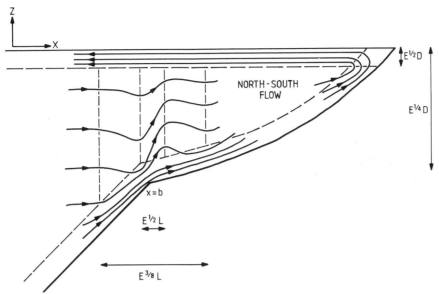

Fig. 11. Cross section showing diagrammatically the circulation streamlines over the shelf break. The shear layer thicknesses are in terms of the Ekman number E [from Johnson and Manja, 1982].

vorticity equation which is then solved numerically. With a shelf-slope topography along the Australian coast, a shelf current is generated on the northwestern shelf and travels southward on the Western Australian shelf. It then veers eastward along the edge of the Great Australian Bight before turning southward off western Tasmania. There is a compensatory countercurrent over the lower slope region and further offshore.

The models described in this section are for homogeneous oceans driven by steady wind stress and therefore lack many observed features. However, they do show that the effect of the rapid change in bottom gradient around the shelf break is a key feature in producing shear regions above the break which may be associated with secondary upwelling and jets along the break. As they represent steady flows, they cannot deal with the spin-up of coastal flows or with upwelling events; these require the variability introduced in the next subsection.

## 3.2. Unsteady Barotropic Models

These models include variability and topography, but by omitting stratification they mostly manage to remain linear theories. The problem of stratification is avoided in some cases by considering only the vertically integrated equations of motion, but these models can then only describe the barotropic component of the flow.

A sequence of papers by Csanady have included shelf break topography and feature in a number of the later subsections. One such paper included here is by Csanady [1974], in which vertically integrated equations are used to examine relative-

ly slow large-scale motions over a shallow shelf and shelf break (or other offshore open boundary). The surface elevation and onshore transport are assumed to be continuous, and the solution is patched to open ocean models through these conditions. The model is spun up from rest at time t = 0 by a suddenly imposed uniform wind, and the discussion is confined to the forced aperiodic motions that are produced. A compound model patches a frictionless "coastal jet" solution in the open ocean to frictionally controlled flow over the shelf. As far as the open ocean is concerned, the continental slope acts almost as a solid wall, with relatively small "leakage" across the shelf edge.

The generation of an alongshore mean flow by a periodic wind stress is studied by Denbo and Allen [1983] in an attempt to develop a model of a poleward eastern boundary undercurrent. The f plane model with a homogeneous interior between surface and bottom Ekman layers is forced by surface wind and restrained by bottom friction. The bottom topography, shown in Figure 12, is the alongshore averaged smoothed depth profile for the Oregon shelf. The traveling applied wind stress is periodic in space and time, with zero time mean. The nonlinear momentum equations are solved using a series expansion, with the Rossby number as a small parameter, and by averaging in time over one wind period. The wind stress curl response is found to be small compared with the coastal forced response. However, the results for coastal forcing are found to be inconsistent with observations of eastern boundary undercurrents, whereas the results for interior forcing are qualitatively consistent with observations. The mean cross-

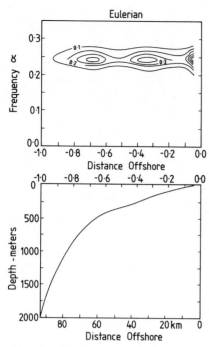

Fig. 12. The depth profile and the Eulerian forced alongshore mean velocity as functions of distance offshore and frequency [from Denbo and Allen, 1983].

shelf flow in the interior and the Ekman layers is driven by gradients of the Reynolds stresses of the oscillatory motion. This net flow must be balanced by a transport in the bottom Ekman layer, which, in turn, necessitates the existence of a mean alongshore velocity. This mean alongshore flow has maxima at distances offshore where there is a rapid change in bottom gradient, as shown in Figure 12 (though such changes are hardly sharp enough to be considered as a "shelf break," the topography here being a smoothed average from the Oregon shelf where the shelf edge is indistinct). This is an interesting new approach that could lead to further developments in the theory, but at the same time it sounds a cautionary note about the relevance of coastal forcing to shelf break activity.

The idea of the shelf break acting as a barrier with only a small amount of leakage across it from shelf to deep water was introduced above. This possibility is investigated further by Johnson and Nurser [1983], who take account of the shelf break by introducing a leaky boundary condition at the edge of the shelf; this allows a weak linkage between the rapidly changing flow on the shelf and the more slowly varying deep water circulation offshore of the shelf break. The scaled gradients of the bottom topography offshore and onshore of the break are $\alpha_L$ and $\alpha_R$, respectively. It is found that the amount of shelf break upwelling ranges from zero when $\alpha_L = \alpha_R$ and the break is absent, through intermediate values when $\alpha_L$ is

moderately larger than $\alpha_R$ and transport leaks off the shelf into the deep ocean, to a maximum as $\alpha_L \rightarrow \infty$ and the shelf break acts as a barrier between the shelf circulation and that in the deep ocean. As the sharpness of the shelf break increases, a stronger current develops along the edge of the shelf.

The effect of increasing the change in slope at the shelf break is illustrated in Figure 13, taken from Johnson and Nurser [1984], which both corrects some numerical values in their previous paper and extends it to a wider range of examples. The increase in the offshore slope $\alpha_L$ causes only a slight increase in the current v along the shelf edge. However, the onshore velocity u is considerably reduced and the shelf break upwelling is considerably increased and comes into phase with u. This shows the tendency, as $\alpha_L$ increases from 2 to 8, of the shelf break to act as an effective barrier as it becomes sharper, and for all onshore flow into the shelf layer to come from the shelf break upwelling. Figure 14 shows an example of this model driven by a wind stress which represents an upwelling event and which is a simplified representation of the wind observed off northwest Africa and shown in Figure 4. The action of the wind stress is confined to the region along the coast between y = −0.25 and y = 0.5. The figure shows that the upwelling at the shelf break is linked to the coastal wind but that the linkage is weak. The largest values of shelf break upwelling occur where the wind is weaker. Also, the longshore current becomes more influenced by the shelf break as the wind stress decreases.

It is shown by Johnson and Nurser [1984] that the net upwelling out of the bottom Ekman layer at the shelf break is given as a function of time by

$$\text{Upwell}(t) = \text{Upwell}(t_0) \exp\{-F_L f(t - t_0)/d\}$$
$$-\int_{t_0}^{t} (F_L - F_R) v_t(t') \exp\{-F_L f(t - t')/d\} dt'$$

where d is the depth of the shelf break; $F_R^2 = (1 + \alpha_R^2)/2f$ $F_L^2 = (1 + \alpha_L^2)/2f$; and $\alpha_R$, $\alpha_L$ are the slopes onshore and offshore of the shelf break, respectively. When $v_t$ is small near the shelf break, Upwell decays exponentially in time with e-folding decay time of $d/fF_L$. Thus more rapid decay is produced when the offshore slope $\alpha_L$, and hence $F_L$, are larger or the depth of the shelf break is smaller. The upwelling is forced by nonzero values of $v_t$ within the recent past (defined in terms of $d/fF_L$). Notice that Upwell is of opposite sign to these recent values of $v_t$ and hence during the buildup of coastal upwelling, when $v_t < 0$, there is secondary upwelling over the shelf break. When the coastal upwelling reaches a quasi-steady state and $v_t$ becomes small, the upwelling over the break decays. These relationships between the wind, the longshore current v, and the function Upwell may be observed in Figures 13 and 14.

Clearly, the inclusion of variability in the

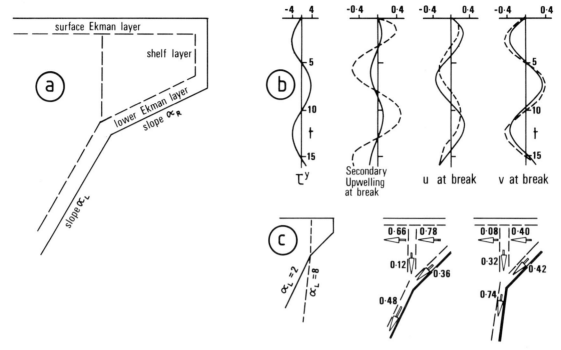

Fig. 13. (a) The shelf region with slope $\alpha_R$ and slope region with slope $\alpha_L$. (b) With $\alpha_R = 1$, $\alpha_L = 2$, the wind stress $\tau^y$, the upwelling at the shelf break, and the onshore (U) and longshore (V) velocities at the shelf break as a function of time t. (c) The transports in the interior regions, the bottom Ekman layer, and the shelf break layer at time t = 7 for two different values of $\alpha_L$ [from Johnson and Nurser, 1984].

models described in this subsection has broadened the range of theoretical methods available and has provided more opportunities to compare theory with observations by the use of either actual wind data or simplified distributions representing certain aspects of the real data, e.g., upwelling events. One distinctive result that emerges from these unsteady barotropic solutions is the fact that the shelf break and a steep slope offshore can act as an effective barrier between the shelf water and the deep ocean circulation, with only limited interaction between the two water masses. In situations such as these a longshore jet may form just above the shelf edge. This leads one to consider the possibility of shelf break fronts.

### 3.3. Models of Shelf Break Fronts

As an interlude, between the descriptions of barotropic and baroclinic models, it is appropriate to discuss here two papers that consider the formation and stability of fronts that may be formed between two different water masses in the neighborhood of the shelf edge.

Flagg and Beardsley [1978] investigate the stability of a Margules front between two immiscible homogeneous layers of different densities, over steep topography. A basic state with steady uniform along-isobath flow in each layer is perturbed by a motion of small amplitude. (This allows the equations of motion to be linearized.) It is found that the effect of increasing the bottom slope is, in general, a reduction of the growth rate of the unstable modes that exist over flat topography. Hence the front may migrate offshore toward the shelf break until an equilibrium is reached between the instability mechanism and local dissipation. This is one possible explanation of why fronts between coastal and oceanic waters are so frequently observed near the transition from shelf to slope, as reported in section 2.

The formation of a surface to bottom front over steep topography is considered by Hsueh and Cushman-Roisin [1983]. They assume a hypothetical barrier between two immiscible homogeneous fluids of different densities, with one fluid confined to the shelf and the other to the deeper region over the slope. The barrier is then removed and geostrophic adjustment permitted. The method of solution is numerical and is based on the assumption of conservation of potential vorticity during adjustment; friction and mixing are assumed to be negligible. The resulting flow is along the front and is in geostrophic balance. Figure 15 shows an example without wind forcing at the surface; the lighter fluid is on the shelf, and the heavier in the deeper region. The shortening of water columns moving up the slope dominates the total tran-

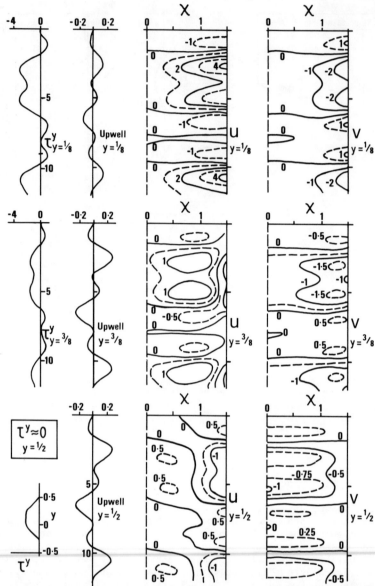

Fig. 14. Sloping shelf with $\alpha_R = 1$, $\alpha_L = 2$. Wind stress $\tau^Y$, upwelling at shelf break (upwell), and onshore (u) and longshore (v) velocities at latitudes y = 1/8, 3/8, and 1/2 as a function of time measured downward [from Johnson and Nurser, 1984].

sport $T_B$ with the generation of a strong lower layer along front flow in the negative y direction. An increase in the steepness of the bottom slope accentuates these gravitational effects. Equilibrium is reached with the front remaining near the shelf break. The examples in which a constant surface wind stress is added should be treated with caution, as the requirement of conservation of potential vorticity is violated.

## 3.4. Steady Baroclinic Models

This set of papers introduces baroclinicity in a variety of ways, including a model with a homo-geneous shelf but stratified deep water, an essentially two-layered model, a well-mixed model in which the density varies only in a horizontal direction, and continuously stratified models that are linearized about a basic density state. An important theoretical problem is how to cope with the nonlinear heat equation; these aforementioned approaches to baroclinicity were constructed to deal with this difficulty.

The first example mentioned above is due to Tomczak and Käse [1974]. It has a homogeneous shallow shelf adjacent to a weakly stratified deep ocean. The model is essentially two dimensional, with no y dependence, and uses the f plane approx-

48    JOHNSON AND ROCKLIFF

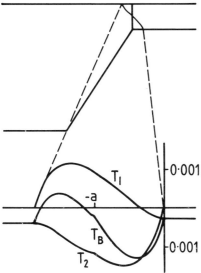

Fig. 15. The shape of the interface for fluids of densities $\rho_1$ and $\rho_2$ ($\rho_2 > \rho_1$) initially separated by a vertical barrier at the shelf break with the lighter fluid on the shelf. Upper layer, lower layer, and total transports are marked $T_1$, $T_2$, and $T_\beta$, respectively [from Hsueh and Cushman-Roisin, 1983].

imation. The aim of the authors was to determine the wind stress driven flow in the deep ocean. The effect of the shallow shelf topography is incorporated by a patching condition at the break in which u and v over the flat shallow shelf are assumed to have Ekman layer solutions driven by a constant wind parallel to the coast. Rather curiously, the bottom Ekman layer on the shelf is not the usual solution, but an artificial inversion of the surface Ekman layer. This condition at the shelf break drives a flow in the deep water where a linearized form of the heat equation is satisfied. As the flow at the break is prescribed, the model can give no information about the circulation dynamics there.

A model which is in many ways the reverse of the previous example is that presented by Hsueh and Ou [1975]. They consider a flat shelf adjacent to a sloping region, and again the effect of the shelf break is represented by a prescribed condition at the shelf edge. This time, however, only the flow on the shelf is calculated. This is a β plane model, which uses a linearized heat equation with the density variation regarded as a small perturbation to a basic linearly stratified state. A prescribed longshore velocity v > 0 at the shelf break induces an upward motion near the bottom and offshore flow near the surface. The general conclusion is that an equatorward longshore current at the shelf break could foster weak secondary upwelling there.

A more interesting two-layer model in which the flow right across the shelf is determined, rather than being prescribed, is set on an f plane by

Heaps [1980]. The interface between the two layers is fixed, and the wind driving is confined to the upper layer. The novel feature of this model is the radical alteration in turbulence conditions at thermocline level that occurs on passing from the shelf to the deeper ocean over the upper part of the continental slope. This change in mixing was inferred from observed isotherm patterns. Heaps assumes that the change in the nature of the turbulent mixing in the thermocline region immediately below the surface leads to an effective rapid increase in the friction coefficient k in the seaward direction over the edge of the continental shelf. The model is two dimensional, with no y dependence. Shelf edge upwelling is shown to be proportional to $(-\tau^x \, \partial k/\partial x)$, where $\tau^x$ is the onshore wind stress (and x increases shoreward). Thus upwelling occurs above the shelf break if the wind is onshore ($\tau^x > 0$) and if k increases offshore ($\partial k/\partial x < 0$). The model is applied to the southwest European continental shelf, and it is thought that the mechanism presented here may in fact operate in that region. Clearly, however, there is room for further development of this theory.

The effects of alongshore variations in bottom topography on a boundary current in a weakly and continuously stratified ocean model on an f plane are described by Janowitz and Pietrafesa [1982]. Here the depth distribution h(x, y) includes a shelf break but also incorporates small variations along the coast (with $h_x \gg h_y$). There is a prescribed current upstream and offshore, representing the Gulf Stream. The density distribution is split into a basic constant barotropic part plus a perturbation baroclinic part. Both the barotropic and the baroclinic solutions are expanded as power series in the small Rossby number. The lowest-order barotropic flow is along the isobaths. Upwelling (downwelling) occurs if the slope decreases (increases) in the downstream direction so that the isobaths diverge (converge). For the baroclinic flow the bottom velocity is along the isobaths, but under upwelling conditions the velocity vector rotates clockwise as the height above the bottom increases. To lowest order, the baroclinic sheared current follows the shelf break zone isobaths, but a variation in the vertical component of relative vorticity along an isobath is produced in regions where the bottom slope varies in the downstream flow direction. This variation in vorticity induces vertical and cross-shelf flows. When the model is applied to the northeast Florida shelf, it is found that the divergence of isobaths to the south of Cape Canaveral causes onshore flow, while the convergence of isobaths further north results in offshore flow. The onshore flow is related to anomalously cool bottom waters observed by Blanton [1971].

The final paper to be described in this subsection forms a link between the sections on steady and unsteady baroclinic models, as it naturally includes a slowly varying density field. Shaw and Csanady [1983] have produced a model of self-ad-

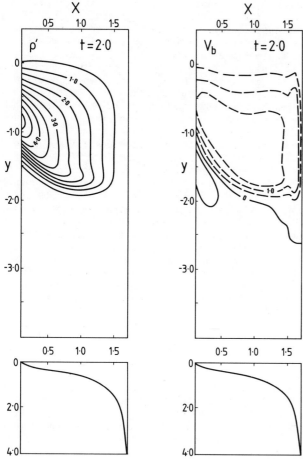

Fig. 16. (a) Contours of density field produced by surface cooling and (b) the corresponding long-shore bottom geostrophic velocity, both at time t = 2. The cooling is applied at -1 < y < 0 from t = 0 to t = 2. Depth variation is shown in lower panel [from Shaw and Csanady, 1983].

tion, which has known analytical solutions. The disadvantage of this method is that self-advection rarely occurs, as the advection by wind-driven currents almost always dominates.

At the shelf break, Shaw and Csanady [1983] apply a "boundary condition" that requires the pressure to remain constant there, so that there is no onshore bottom flow, representing the "barrier" effect of the steeper slope beyond, as discussed in section 3.2. A prescribed cooling is applied to the surface, and then the resulting distributions of the density and velocity fields are determined numerically. Figure 16 shows the alongshore bottom geostrophic velocity and the well-mixed density field produced by the surface cooling

$$Q(y) = (\pi/2) \sin \pi y \qquad -1 < y < 0$$

and Q zero elsewhere. The strong flow exhibited at the shelf break is somewhat unrealistic, as the boundary condition enforced there is approximate and because the assumption of vertical homogeneity breaks down. Note the formation of a density front between y = -1 and y = -2 as the density perturbation is advected in the negative y direction. A number of other cases are described in the paper, with the general conclusion that a dense water blob is advected with the shallow water on the right in the northern hemisphere, whereas the direction of propagation of a light water blob is reversed. A strong density front may appear in the forward face of the density perturbation.

### 3.5. Unsteady Baroclinic Models

From the remarks made in the introduction to section 3 it is clear that the models in this section should be most able to reproduce observed results, as they have the three fundamental requirements of topography, stratification, and variability. However, it should become apparent that there is still much further research to be carried out to improve the agreement between observation and experiment by refining the models.

An early model that incorporated stratification by using two layers of different density was that presented by Defant [1952]; it demonstrates the onset of upwelling over a seafloor step parallel to the coast. For a wide shelf of depth H and for a very deep ocean beyond the shelf break, the shelf edge upwelling is the fraction $h_1/\{h_2 + (h_2 H)^{1/2}\}$ of the coastal upwelling, where $h_1$ and $h_2$ are the thicknesses of the upper and lower layers, respectively, over the shelf. The upwelling, which decays on either side of the edge of the step within the appropriate internal deformation radius, is depicted by a rise in the interface between the two layers. The change in depth at the shelf edge causes divergence in the upper layer, and shelf break upwelling results. The shelf has to be sufficiently wide for the influence of the coast to be separate from the in-

vection of density perturbations on a sloping continental shelf. The model, on an f plane, has a long straight shelf with depth h(x). As density perturbations in the model are expected to be caused by local surface cooling and overturning, it is assumed that the density field is well mixed in the vertical, so that $\rho = \rho(x, y)$. The momentum equations are a balance between geostrophy and bottom friction, with the time-dependent terms assumed to be negligible. However, in the heat equation the unsteady $\partial \theta / \partial t$ term is retained to allow changes on a slow time scale brought about by advection. The authors attempt to justify this inconsistency by claiming that in the momentum equations the cross-isobath Coriolis acceleration and the along-isobath bottom friction are relatively large. The mathematical advantage that arises from this choice of dominant terms is that away from boundary layers the self-advection of the density field is governed by Burgers' equa-

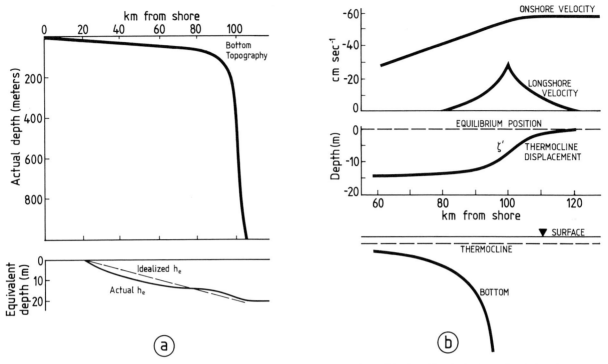

Fig. 17. (a) Depth distribution over continental shelf off the Virginia-Maryland border and the "equivalent" depth. (b) "Shelf jet" and thermocline displacement generated by a shore-parallel wind [from Csanady, 1973].

fluence of the edge (that is, at least two deformation radii wide).

Another two-layer model which allows more realistic bottom topography is that of Csanady [1973]. An equivalent depth $h_e = h_1 h_2/(h_1 + h_2)$ is introduced to smooth out the sharp bottom topography, as shown in Figure 17, with $h_1$ and $h_2(x)$ again the upper and lower layer thicknesses. The depth $h_e$ is a slowly varying function of position, and the rapid depth variations at the edge of the shelf hardly affect the $h_e$ distribution, as $h_2 \gg h_1$ there and therefore $h_e \approx h_1$. The solution is determined analytically in terms of Bessel functions for various steady uniform wind stresses applied impulsively at $t = 0$. Only the forced response is calculated, and no account is taken of the free waves generated by the impulsive wind stress. In the solution the coastal upwelling remains separate from its shelf break analogue so long as the shelf width is reasonably large. For an onshore wind the associated thermocline displacement contains a steplike feature near the edge of the shelf and assumes a constant depression closer to the shore. On the other hand, as shown in Figure 17, for a longshore wind, displacements are much larger (and grow linearly with time, so that eventually the linear theory breaks down), although the thermocline slope is similar to the first case. The shelf break jet accompanying these thermocline displacements is of modest amplitude compared with other theoretically predicted coastal jets. The cusp is due to the discontinuity in the chosen $h_e$ and would be smoothed for any real topography. These results are only valid for the special case with no dependence on the y coordinate. As pointed out by Allen [1975], this model cannot be generalized to include alongshore gradients, as in this more realistic situation the equations for the baroclinic mode may not be derived independently of the barotropic component.

A two-dimensional f plane model which is independent of the y coordinate but includes weak stratification and variability is introduced by Janowitz and Pietrafesa [1980]. All the fields are expanded as power series in the stratification parameter s, with $s \ll 1$. First, the barotropic lowest-order contribution, in which the density is regarded as a dynamically passive scalar and in which the bottom friction can be either linear or nonlinear, is determined. Then the changes in the density field induced by the barotropic velocities are calculated. Finally, the second-order corrections to the velocity fields as driven by the lowest-order density field are determined. It is found that for a constant wind stress $\tau^y$ for $t > 0$ the shallow water spins up first, producing a coastal jet and an upward bulge in the isopycnals which propagates offshore. At the shelf break, if $hh_{xx}/h^2_x > 2$, another upward bulge is formed in the isopycnals, indicating shelf break upwelling; this is of larger amplitudes than the propagating

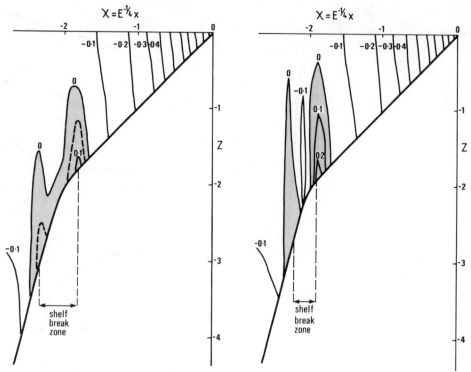

Fig. 18. Contours of longshore velocity in a cross-shelf section: (a) for a relatively broad shelf break zone and (b) for a narrower shelf break region, showing two poleward undercurrents with an intensified equatorward flow between.

bulge. The theory is compared with current meter data collected in Onslow Bay, North Carolina. This is a very simplified theory which incorporates many physical effects in a neat way. It would be very useful if it could be extended to a three-dimensional case while retaining much of its analytical simplicity.

The final paper described in this section is a two-time-scale model of the circulation over the shelf break. Johnson [1982] argued that barotropic changes over the shelf occur on a short time scale of hours to days, whereas baroclinic changes take days to weeks to spin up. Therefore a suitable expression for the pressure p is

$$p = p*(x, y, t*) + \hat{p}(x, y, z, \hat{t})$$

where the barotropic part p* is independent of depth z and changes rapidly, on the barotropic time scale, and the baroclinic part $\hat{p}$ does vary with z and changes slowly, on the baroclinic time scale. A similar splitting into barotropic and baroclinic parts is used for the velocity field, but the density (or temperature) field is purely baroclinic. The momentum equations contain the unsteady $\partial v*/\partial t*$ term, and the heat equation contains the slowly varying $\partial \hat{\theta}/\partial \hat{t}$ term, so that full account is taken of variability. The theory is developed for p* and $\hat{p}$ of similar magnitudes, allowing adequate influence of baroclinicity. The

method of calculation involves a numerical procedure that first finds the barotropic components driven by an imposed variable wind stress. Then the slowly varying temperature field produced by advection with the flow is calculated; this is then used to find the baroclinic components of the pressure and velocity fields. The procedure continues by recalculating the barotropic components and so on.

N. Rockliff (unpublished manuscript, 1985) has applied this two-time-scale method to the flow over a smooth but intense shelf break. The shelf-slope topography depicted in Figure 18 is modeled by straight sloping shelf and slope regions, linked by a smooth curve, with continuity at the matching points of up to the second derivative of the depth function. An initial imposed density field is first allowed to adjust in the absence of forcing, and then a surface wind stress $\tau^y$, periodic in the longshore direction and varying with time to simulate an upwelling event followed by downwelling, is applied. The results show a number of important features, some of which are also predicted by other models, and some of which reflect observed features. The first of these is the backlog of onshore flow by the topography: as the gradient of the continental, slope region is increased, the onshore velocities are decreased. This is similar to the predictions of the barotropic model of Johnson and Nurser [1983].

The second is the appearance, during upwelling
favorable winds, of a poleward undercurrent situated over the shelf break zone. This current,
which forms at the bottom, appears to be most
intense at the upper edge of the shelf break zone,
with a secondary maximum at the lower edge, where
the continental slope proper begins. In fact, it
may initially appear as two separate jets, which
may or may not merge into one over the whole shelf
break zone, depending on the width of the zone, on
the relative steepnesses of shelf and slope, and
on the duration of the upwelling favorable
winds. Figure 18a shows contours of longshore
velocity in a cross section, for a relatively
broad shelf break zone and a moderately steep
continental slope region. Increasing the gradient
of the continental slope appears to intensify the
currents and increase their vertical extent, as
well as induce their earlier formation. A narrow-
ing of the shelf break zone also increases current
velocities but appears to make more likely the
formation of two poleward jets separated by a thin
region of intensified equatorward flow. This
configuration, illustrated in Figure 18b, bears
some resemblance to the intense equatorward jet
structure observed by Bang and Andrews [1974] (see
Figure 6) and also to results of the numerical
model of Hurlburt and Thompson [1973]. Finally,
the isotherms (Figure 19) show evidence of upward
water movements in the region of the shelf break,
as well as normal coastal upwelling.

## 3.6. Layered Numerical Models

Fully numerical models that obtain solutions to
the primitive equations of motion using finite
differences have various ways of dealing with the
vertical stratification. The most common methods
are discrete layers or distinct levels or the use
of eigenfunctions and modes. The first and last
methods are discussed in this and the next subsections. Models based on many vertical levels are
not usually used for shelf dynamics but are more
common for large-scale ocean modeling.

A series of papers describing two-layer models
have been presented by Hurlburt, O'Brien and
Thompson. The first paper in this series is by
O'Brien and Hurlburt [1972] and presents a two-
dimensional model independent of y and based on an
f plane. Nonlinear momentum terms and horizontal
friction terms are retained in the equations of
motion in addition to the usual unsteady, geostrophic, and vertical friction terms. A semi-impli-
cit scheme is used to solve for the velocity components and the thicknesses of the two layers
driven by a constant uniform wind stress for $t >$
0. With a sharp shelf break, a weak secondary
upwelling manifests itself above the shelf edge as
a displacement of the interface between the two
layers. Some shear in the longshore flow occurs
at the break. A similar numerical experiment
using topography that simulates the Newport, Oregon, shelf (with a very weak shelf break) does not
produce any noticeable effect on the shelf edge.

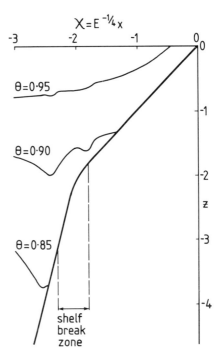

Fig. 19. Isotherms in a cross-shelf section,
showing coastal upwelling and possible secondary
upwelling at the shelf break.

The above model is extended by Hurlburt and
Thompson [1973] to include the β effect and a
longshore pressure gradient but retains the condition that the velocity field is independent of
y. The numerical scheme is similar and calculates
the flow induced by a constant uniform wind
stress. The same sharp shelf break topography is
used, and again a secondary upwelling zone occurs
over the shelf edge in the form of an upward bowing of the interface, as shown in Figure 20, nearly 100 km offshore. Strong shear in the longshore
current occurs around the edge of the shelf, but
there is no indication of any longshore jet.
During the spin-up phase, when $v_t < 0$, the offshore transport over the shelf is slightly weaker
than that over the slope. As a result, a secondary upwelling zone develops over the slope. The
growth in secondary upwelling ceases as $v_t \to 0$.
This is similar to the theory of shelf break upwelling of Johnson and Nurser [1984]. Secondary
upwelling was more likely in β plane cases than in
f plane cases. It is claimed that the effects of
bottom topography depend on the effective depth of
the Sverdrup balance. As the ocean depth becomes
shallower than this depth, the onshore flow becomes supergeostrophic. This results in a tempor-
ary equatorward acceleration, eventual frictional
balance, and the formation of a secondary upwelling zone over points of large bottom slope such as
just offshore of the edge of the shelf.

The third paper in this series, by Thompson and
O'Brien [1973], looks at the effect of a time-
dependent wind on the f plane two-layer model of

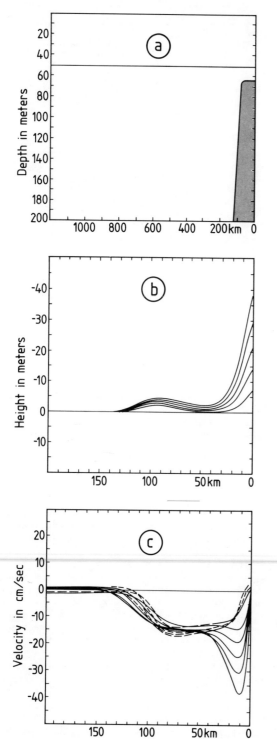

Fig. 20. (a) Bottom topography, (b) pycnocline height anomaly, and (c) longshore velocity at 5-day intervals up to day 25 [from Hurlburt and Thompson, 1973].

O'Brien and Hurlburt [1972] described above. The applied surface wind has longshore oscillations whose kinetic energy spreads over a wide spectral band from $10^{-3}$ to $10^{0}$ cycles per hour. A secondary upwelling zone forms over the shelf break with the same width scale (30 km) as the sloping shelf edge. Increasing the bottom slope increases the maximum height anomaly of the secondary upwelling zone and decreases its width. Secondary upwelling is a transient feature which decays away under a steady wind stress and is explained in terms of the potential vorticity

$$\frac{\partial v_i}{\partial x} = \frac{f(h_i - H_i)}{H_i} \qquad i = 1, 2$$

where the $H_i$ are the initial thicknesses of the upper and lower layers, respectively, and the $h_i$ and $v_i$ their thicknesses at time t and longshore velocities, respectively. At the shelf edge, $v_1$ decreases shoreward under northward winds (similar to the example in Figure 20); hence $h_1 < H_1$ and upwelling is observed.

A more recent paper which describes a numerical model of coastal upwelling off Peru and which includes a mixed layer is by O'Brien and Heburn [1983]. They consider a two-layer vertically integrated primitive equation model on a β plane. The effects of bottom topography are restricted to the lower layer, with only weak coupling to the upper layer, since observations off Peru suggest that the lower layer is isolated from local surface forcing. The model is forced by local winds and by externally excited coastal-trapped waves. The bottom topography is digitized from the Peru shelf, with a broad shelf (about 20 km) in the north and a narrow shelf (about 6 km) in the south, with variable shelf slope in the longshore, as well as the offshore, direction. There is entrainment between the upper and lower layers, which are both well mixed in the vertical. The local wind forcing is based on low-pass-filtered observations from a mooring near 15°S, retaining periods greater than 12 hours to include simulation of the land/sea breeze cycle. The results of the model do not concentrate on shelf break effects, but Figure 21 shows a typical cross section for v and a sea surface temperature field with coastline and bottom topography. In the temperature field a warm plume has formed in the region above the shelf edge, with a cooler region just offshore; there is also cooler water at the coast due to coastal upwelling. The cool water is associated with the interaction of the imposed internal wave forcing with the ridge in the bottom topography located in this region and with enhanced upwelling there.

To conclude this subsection on layered numerical models, there follows a brief review of two papers, by Fandry [1982, 1983], which describe models of the Bass Strait between mainland Australia and Tasmania. The earlier paper uses a single-layer vertically integrated linear numerical

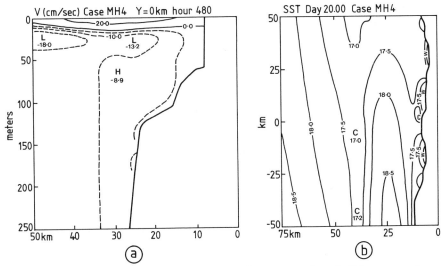

Fig. 21. (a) Cross-shelf velocity and topography at y = 0 and (b) sea surface temperature field and coastline slope, both at model day 20 [from O'Brien and Heburn, 1983].

model including bottom topography on an f plane. The topographic effect is dominant in producing barotropic currents along the depth contours, especially around the shallow edge of the eastern continental shelf. The isolines of the sea surface height nearly coincide with the isobaths in this region. The later paper combines the dynamics of the first model with an analytical solution of the Ekman equations, which at each grid point provide an expression for the time-dependent flow at any depth in terms of a convolution integral over the sea surface slope and the wind stress. Again the main features of the solution are a tendency for contours of surface elevation to align themselves with the isobaths, with the bottom topography being a dominant influence on the flow.

3.7. Continuously Stratified Numerical Models

This final collection of numerical models adopts a variety of ways of dealing with the vertical variations of the density and velocity fields, including spectral methods involving eigenfunction expansions, methods and variable grid spacings, and straightforward use of the heat equation to determine changes in density.

In a paper that is slightly peripheral to the topic of this review, Davies and Heaps [1980] present a three-dimensional model of the Norwegian Trench in the North Sea basin using a spectral method. The bottom topography consists of a deep trench adjacent to a shallow shelf representing the North Sea. There is inflow at nearly all depths of the shallow North Sea (reaching a maximum near the edge of the shelf) and outflow through most of the trench. There is a large amount of shear above the shelf edge. The vertical variation is dealt with by assuming that the

vertical eddy viscosity is proportional to the depth h and by representing the velocity fields by a truncated Fourier series.

Continuing with this spectral method, a recent paper by Heaps and Jones [1985] describes a three-layer model of wind-induced motion over a bottom slope, with the stratification in each layer represented as an expansion of eigenfunctions. The bottom topography h = h(x) has three parts, consisting of a deep, flat ocean, a slope region with constant gradient, and a shallow, flat shelf. The flow is driven by a wind parallel to the shelf edge over a limited distance along the shelf. It is found that strong upwelling occurs near the top of the slope in the region of wind forcing. Inertial currents, largest in the shallower water, are also excited. Figure 22 shows the rise in the interfaces between the layers over the slope, with the maximum elevation occurring at the shelf edge. In the upper layer a shelf edge jet is formed ($v_1$). The variations in w shown in Figure 23 indicate an undulating flow of some complexity over the slope. The length of the wind region and the magnitude and direction of the wind forcing are important parameters governing wave generation. This is a promising new model and is expected to be developed further to investigate the effect of changing the wind forcing, the frictional parameters, the simulation period, the stratification, and the boundary conditions. A more realistic shelf edge slope needs to be considered.

A continuously stratified numerical model in which changes in density are determined from a finite difference form of the heat equation is presented by Hamilton and Rattray [1978]. The model on the f plane is wind driven, for periods of up to 10 days, and in some examples includes a shelf break, with rectangular topography offshore of the shelf edge. Although a longshore pressure

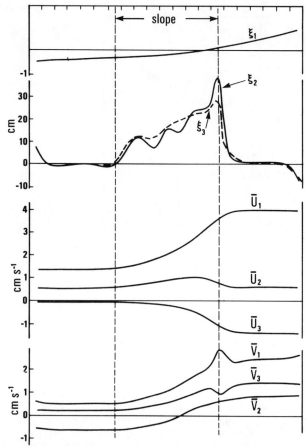

Fig. 22. Distribution of surface elevation $\xi_1$, interface elevations $\xi_2$, $\xi_3$, and depth mean currents $(\bar{u}, \bar{v})$ in the three layers for a section across the slope at t = 8 hours [from Heaps and Jones, 1985].

gradient can exist on the seasonal time scale, it is claimed that its effect is small over the shelf and that it may be neglected when the variability of the upwelling is considered over much shorter periods. Consequently, this model has no variation of quantities in the longshore direction. Figure 24 shows that the denser water on the bottom at midshelf has been advected from below the shelf break. Data from Barton et al. [1975] show that the upwelled water at midshelf is not directly connected with water of the same density seaward of the break. This water is apparently upwelled at some position to the north of the section and advected southward by the longshore flow. Thus the model produces a two-dimensional representation of this three-dimensional effect, but of course what is needed really is a truly three-dimensional model. The transverse velocity distribution in Figure 24 shows evidence of secondary upwelling at the shelf break associated with the abrupt change in gradient.

The practical limitations of the two-dimensional model of Hamilton and Rattray [1978] are investigated by Hickey and Hamilton [1980], who compare the model with observations on the Oregon-Washington shelf. They use the model to predict forward by up to 31 days the isopycnal distribution at Westport, Washington. The model appears to be effective for time periods as long as 15 days for the prediction of offshore displacements of surface isopycnals and of vertical displacement of isopycnals below the surface layer. Better results were obtained using a wind station 200 km to the south rather than a wind station only 50 km to the south, thus demonstrating the importance of free waves traveling along the coast. The model should not be used over long periods without supporting data to test it at frequent intervals.

Another continuously stratified model in which the initial stratification consists of two homo-

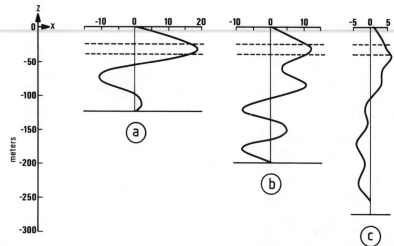

Fig. 23. Vertical profiles of upward velocity w at stations (a) on the shelf, (b) on the slope, (c) in the deepwater, all at t = 8 hours (adapted from Heaps and Jones [1985]).

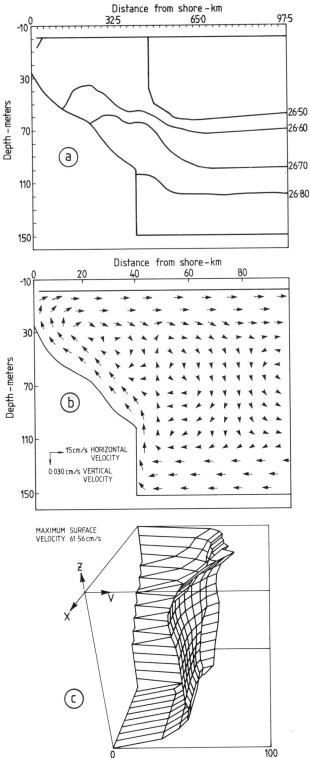

ELAPSED TIME - 9 DAYS

Fig. 24. (a) Isopycnal distribution, (b) cross-shelf velocity u, and (c) longshore velocity v, all after 9 days of wind [from Hamilton and Rattray, 1978].

geneous layers of different density separated by a pycnocline region, in which the density varies linearly with depth, is derived by Suginohara [1982]. At later times, the density distribution changes due to advection and diffusion through the heat equation. An f plane is used, and the shelf topography is based on that off Oregon, so that the shelf break is weak. There is some evidence in the results of the numerical calculation of enhanced longshore currents over the broad shelf break zone, and the deviations of density from the initial state are strongest there.

In an exposition of a two-mode numerical model with applications to coastal upwelling, Van Foreest and Brundrit [1982] describe a continuously stratified β plane model which includes both the barotropic and the first baroclinic modes coupled by topography. The vertical dependence in the equations is removed by using a Galerkin procedure which has the normal modes as test functions. The model is forced by a body force $\tau^{x,y} Z(z)$, where $Z(z)$ is zero except in the upper mixed layer. The first experiments are two dimensional, to compare the two-modal solution with the two-layer solution of Hurlburt and Thompson [1973] described above. As shown in Figure 25, for sharp shelf topography the density levels in these two-dimensional experiments rise as the shelf break is approached, drop at the shelf break, and rise again at the coast, indicating shelf break upwelling. A strong long-shelf jet is formed at the shelf break as well as at the coast. The calculations are repeated for a three-dimensional case, with smoothed topography from the southwest coast of South Africa, where, for example (see Figure 25), the maximum longshore flow occurs between the canyon and the break on the Cape Point section.

A numerical model which incorporates the bottom topography and shape between Cap Blanc and Cap Vert in northwest Africa is by H. Meier-Fritsch (unpublished manuscript, 1983). The model adopts a varible grid on the f plane with a fully explicit numerical scheme for the nonlinear stratified equations of motion. He shows that the inclusion of vertical diffusion in the heat equation has little effect on the isopycnal distribution and that this distribution is mainly determined by advection. It is found that increasing meridional winds lead to enhanced horizontal and vertical water movements. Variations in the initial vertical density distributions do not appreciably affect the typical density features. Model results show upward bowing of isopycnals near the shelf edge.

4. Concluding Remarks

These final paragraphs briefly summarize some of the important results from the preceding sections and pose various quetions about how future work on shelf edge dynamics might proceed. In section 2, on observations, confining our interest to currents and density fields rather than wave motions, the most common properties of shelf edge

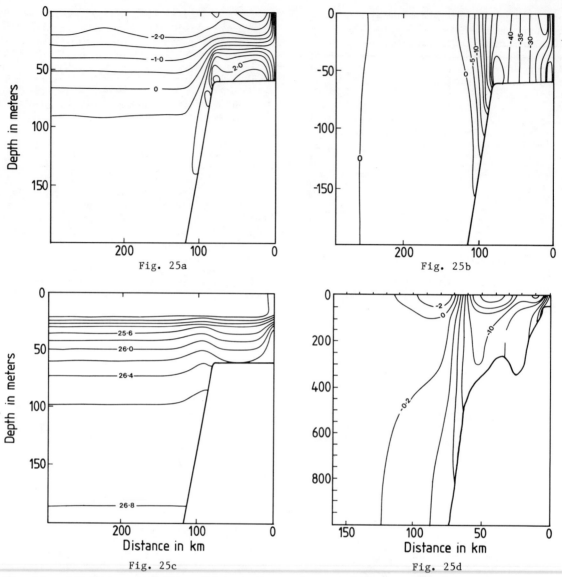

Fig. 25a

Fig. 25b

Fig. 25c

Fig. 25d

Fig. 25. (a) Onshore velocity, (b) longshore velocity, (c) sigma-t section, all at day 10 for the sharp shelf two-dimensional case, and (d) Cape Point section of longshore velocity at day 3 for three-dimensional case [from Van Foreest and Brundrit, 1982].

flows are intermittent secondary upwelling, usually shear in the velocity field, sometimes jets along the shelf break, and, in many places, density fronts between the shelf water and the deep ocean water. How well can these observations be described by the various theories?

A wide variety of analytical and numerical models of varying degrees of complexity have been described in section 3 with a gradual inclusion of more realistic data. Depending on the features included in each model, they can, as a group, broadly describe the observations mentioned in the last paragraph, but the use of real wind data and real topography and coastline shape is not wide-

spread. The basic requirement of models to include topography, stratification, and variability is very difficult because of the nonlinear equation of heat. Consequently, most models have to compromise and treat at least one of these three requirements in a simplified way. Other features whose omission has often been regretted by many authors include the lack of realistic coastline and shelf break shape, the poor matching to the deep water circulation, and a oversimplified choice of wind driving. The two latter properties need more input from observations.

From the observational side, is it possible when designing shelf experiments to include arrays that span the shelf break and resolve the shear and jets that sometimes occur above the shelf edge? Can measurements be taken along the shelf edge to detect three-dimensional effects? Better observations of winds are required for input into the models with sufficient detail to define the wind field both across the shelf and the break and along the shelf. Is it feasible to devise more laboratory experiments, like those described by Whitehead [1981], to understand further the physical mechanisms involved in the shelf break circulation?

Are there straightforward ways of extending some of the models described in section 3 to account for more physical features? Simplified models, such as those of Janowitz and Pietrafesa [1980], are often most useful in elucidating the physical processes involved in a complex flow. Can such a model be extended to the three-dimensional case by inclusion of dependence on y without losing its chief merit of simplicity? The interesting new approach of Denbo and Allen [1983], using periodic wind forcing and interaction with offshore waves, should be developed further. They sound a cautionary warning to modelers that coastal wind forcing may not be particularly relevant to shelf break activity and that more extensive wind distributions should be used. The two-time-scale approach of Johnson [1982] will be extended to include smoothed coastline shape and bottom topography and to be driven by more general wind forcing.

The simplified model of Heaps [1980] introduces a new concept to shelf break modeling in which changes in turbulence conditions at the shelf edge were parameterized by changes in eddy viscosity. Can this theory be developed to include Coriolis acceleration and an active lower layer? This concept of changing eddy viscosity is followed by Heaps and Jones [1985] with the inclusion of three layers with different eddy viscosities but without any lateral changes at the shelf break. A development of this three-layer numerical model to include more realistic wind distributions and more natural bottom topography would be most welcome. The use of Lagrangian as well as Eulerian velocities in numerical models could help with the interpretation of data obtained from floats.

In general, the most important advance that is needed in models of shelf break activity is a better understanding and hence a better representation of the coupling between the shelf circulation and the deep water circulation. Most models need either to incorporate more precise matching conditions between their shelf flows and more realistic deep water flow or to determine more appropriate boundary conditions to apply at their seaward edges. Large time series observations that extend beyond the shelf break and along the shelf edge should help with understanding this coupling.

Another poorly understood property of the circulation near the shelf break is the formation and maintenance of density fronts over the shelf edge. Models are needed that explain why the front forms near the shelf edge and why it is stable in that position. Most existing models set the front at the shelf break at t = 0 and investigate its later properties. Such models cannot explain why the front forms in this neighborhood. It would also be interesting to investigate the role of runoff and of other density-driven flows over the shelf.

The ultimate test of circulation models is their use to predict the flow ahead in time from a set of data at the current time. This has been tried by Hickey and Hamilton [1980], and useful predictions were made over periods of 10 days for comparison with data. They found that the accuracy for long periods was conditional on suitable data being available to update the calculations. The development of models capable of accurate prediction will depend on sufficiently long data sets being available for the testing stage of the model. Clearly, there is still much to do in observational and theoretical work; by combining the two some useful and important results could be obtained.

References

Allen, J. S., Coastal trapped waves in a stratified ocean, J. Phys. Oceanogr., 5, 300-325, 1975.

Allen, J. S., Models of wind-driven currents on the continental shelf, Annu. Rev. Fluid Mech., 12, 389-433, 1980.

Allen, J. S., R. C. Beardsley, J. O. Blanton, W. C. Boicourt, B. Butman, L. K. Coachman, A. Huyer, T. H. Kinder, T. C. Royer, J. D. Schumacher, R. L. Smith, W. Sturges, and C. D. Winant, Physical oceanography of the continental shelves, Rev. Geophys., 21, 1149-1181, 1983.

Andrews, J. C., Thermal waves on the Queensland shelf, Aust. J. Mar. Freshwater Res., 34, 81-96, 1983.

Bang, N. D., The southern Benguela current region in February, 1966, II, Bathythermography and air-sea interactions, Deep Sea Res., 18, 209-224, 1971.

Bang, N. D., and W. R. H. Andrews, Direct current measurements of a shelf-edge frontal jet in the southern Benguela system, J. Mar. Res., 32, 407-419, 1974.

Barton, E. D., and M. L. Argote, Hydrographic variability in an upwelling area off northern Baja California in June 1976, J. Mar. Res., 38, 631-649, 1980.

Barton, E. D., R. D. Pillsbury, and R. L. Smith, A compendium of physical observations from JOINT-I: Vertical sections of temperature, salinity and sigma-t from R. V. Gillis data and low pass filtered measurements of wind and currents, Ref. 75-17, 60 pp., School of Oceanogr., Oreg. State Univ., Corvallis, 1975.

Barton, E. D., A. Huyer, and R. L. Smith, Temporal variation observed in the hydrographic regime near Cabo Coveiro in the northwest African upwelling region, February to April 1974, Deep Sea Res., 24, 7-23, 1977.

Blanton, J., Exchange of Gulf Stream Water with North Carolina Shelf Water in Onslow Bay during stratified conditions, Deep Sea Res., 18, 167-178, 1971.

Blanton, J. O., L. P. Atkinson, L. J. Pietrafesa, and T. N. Lee, The intrusion of Gulf Stream Water across the continental shelf due to topographically-induced upwelling, Deep Sea Res, 28, 393-405, 1981.

Bowden, K. F., Summing-up, Philos. Trans. R. Soc. London, Ser.A, 302, 683-689, 1981.

Brink, K. H., D. Halpern, and R. L. Smith, Circulation in the Peruvian upwelling system near 15°S, J. Geophys. Res., 85(C7), 4036-4048, 1980.

Brink, K. H., D. Halpern, A. Huyer, and R. L. Smith, The physical environment of the Peruvian upwelling system, Prog. Oceangr., 12, 285-305, 1983.

Brockmann, C., E. Fahrbach, A. Huyer, and R. L. Smith, The poleward undercurrent along the Peru coast: 5 to 15°S, Deep Sea Res., 27, 847-856, 1980.

Cannon, G. A., N. P. Laird, and T. V. Ryan, Flow along the continental slope off Washington, autumn 1971, J. Mar. Res., suppl., 97-107, 1975.

Csanady, G. T., Wind-induced baroclinic motions at the edge of the continental shelf, J. Phys. Oceanogr., 3, 274-279, 1973.

Csanady, G. T., Barotropic currents over the continental shelf, J. Phys. Oceanogr., 4, 357-371, 1974.

Davies, A. M., and N. S. Heaps, Influence of the Norwegian Trench on the wind-driven circulation of the North Sea, Tellus, 32, 164-175, 1980.

Defant, A., Das Kaltwasserauftriebsgebeit vor der Küste Südwestafrikas, Ländeskundliche Forsch., Festschr. N. Krebs, 52-66, 1936.

Defant, A., Theoretische Überlegungen zum Phänomen des Windstaus und des Auftriebes an ozeanischen Küsten, Dtsch. Hydrogr. Z, V, 69-80, 1952.

Denbo, D. W., and J. S. Allen, Mean flow generation on a continental margin by periodic wave forcing, J. Phys. Oceanogr., 13, 78-92, 1983.

Dickson, R. R., P. A. Gurbutt, and V. N. Pillai, Satellite evidence of enhanced upwelling along the European continental slope, J. Phys. Oceanogr., 10, 813-819, 1980.

Fandry, C. B., A numerical model of the wind-driven transient motion in Bass Strait, J. Geophys. Res., 87(C1), 499-517, 1982.

Fandry, C. B., Model for the three-dimensional structure of wind-driven and tidal circulation in Bass Strait, Aust. J. Mar. Freshwater Res., 34, 121-141, 1983.

Flagg, C. N., and R. C. Beardsley, On the stability of the shelf water/slope water front south of New England, J. Geophys. Res., 83(C9), 4623-4631, 1978.

Freeland, H. J., W. R. Crawford, and R. E. Thomson, Currents along the Pacific coast of Canada, Atmos. Ocean, 22, 151-172, 1984.

Freeland, L. R., and K. L. Denman, A topographically controlled upwelling center off southern Vancouver Island, J. Mar. Res., 40, 1069-1093, 1983.

Hagen, E., Mesoscale upwelling variations off the west African coast, in Coastal Upwelling, Coastal Estuarine Sci. Ser., vol. 1, edited by F. A. Richards, pp. 72-78, AGU, Washington, D. C., 1981.

Halliwell, G. R., and C. N. K. Mooers, The space-time structure and variability of the shelf water-slope water and Gulf Stream surface temperature fronts and associated warm core eddies, J. Geophys. Res., 84(C12), 7707-7725, 1979.

Hamilton, P., and M. Rattray, A numerical model of the depth-dependent, wind-driven upwelling circulation on a continental shelf, J. Phys. Oceanogr., 8, 437-457, 1978.

Hart, T. J., and R. I. Currie, The Benguela current, Discovery Rep., 33, 123-298, 1960.

Heaps, N. S., A mechanism for local upwelling along the European continental slope, Oceanol. Acta, 3, 449-454, 1980.

Heaps, N. S., and J. E. Jones, A three-layered spectral model with application to wind-induced motion in the presence of stratification and a bottom slope, Cont. Shelf Res., 4, 279-319, 1985.

Hickey, B. M., The California current system--Hypotheses and facts, Prog. Oceanogr., 8, 191-279, 1979.

Hickey, B. M., and P. Hamilton, A spin-up model as a diagnostic tool for interpretation of current and density measurements on the continental shelf of the Pacific Northwest, J. Phys. Oceanogr., 10, 12-24, 1980.

Hill, R. B., and J. A. Johnson, A theory of upwelling over the shelf break, J. Phys. Oceanogr., 4, 19-26, 1974.

Hsueh, Y., and B. Cushman-Roisin, On the formation of surface to bottom fronts over steep topography, J. Geophys. Res., 88(C1), 743-750, 1983.

Hsueh, Y., and H.-W. Ou, On the possibilities of coastal, mid-shelf and shelf break upwelling, J. Phys. Oceanogr., 5, 670-682, 1975.

Hsueh, Y., G. O. Marmorino, and L. L. Vansant, Numerical model studies of the winter storm response of the west Florida shelf, J. Phys. Oceanogr., 12, 1037-1050, 1982.

Hughes, P., and E. D. Barton, Stratification and water mass structure in the upwelling area off northwest Africa in April/May 1969, Deep Sea Res., 21, 611-628, 1974.

Hurlburt, H. E., and J. D. Thompson, Coastal upwelling on a β plane, J. Phys. Oceanogr., 3, 16-32, 1973.

Huthnance, J. M., Waves and currents near the continental shelf edge, Prog. Oceanogr., 10, 193-226, 1981.

Huthnance, J. M., L. A. Mysak, and D.-P. Wang, Coastal trapped waves, this volume.

Huyer, A., A comparison of upwelling events in two locations: Oregon and northwest Africa, J. Mar. Res., 34, 531-546, 1976.

Huyer, A., and R. L. Smith, A subsurface ribbon of cool water over the continental shelf off Oregon, J. Phys. Oceanogr., 4, 381-391, 1974.

Huyer, A., R. D. Pillsbury, and R. L. Smith, Seasonal variation of the alongshore velocity field over the continental shelf off Oregon, Limnol. Oceanogr., 20, 90-95, 1975.

Huyer, A., R. L. Smith, and B. M. Hickey, Observations of a warm-core eddy off Oregon, January to March 1978, Deep Sea Res., 31, 97-117, 1984.

Ikeda, M., and W. J. Emery, A continental shelf upwelling event off Vancouver Island as revealed by satellite infrared imagery, J. Mar. Res., 42, 303-317, 1984.

Janowitz, G. S., and L. J. Pietrafesa, A model and observations of time-dependent upwelling over the mid-shelf and slope, J. Phys. Oceanogr., 10, 1574-1583, 1980.

Janowitz, G. S., and L. J. Pietrafesa, The effects of alongshore variation in bottom topography on a boundary current (topographically induced upwelling), Cont. Shelf Res., 1, 123-141, 1982.

Johnson, D. R., Current profiles in the Canary current upwelling region near Cap Blanc, March and April 1974, J. Geophys. Res., 81, 6429-6439, 1976.

Johnson, D. R., and W. R. Johnson, Vertical and cross-shelf flow in the coastal upwelling region of Oregon, Deep Sea Res., 26, 399-408, 1979.

Johnson, D. R., T. Fonseca, and H. Sievers, Upwelling in the Humboldt coastal current near Valparaiso, Chile, J. Mar. Res., 38, 1-16, 1980.

Johnson, J. A., A two time-scale model of stratified shelf currents, Cont. Shelf Res., 1, 143-157, 1982.

Johnson, J. A., and P. D. Killworth, A bottom current along the shelf break, J. Phys. Oceanogr., 5, 185-188, 1975.

Johnson, J. A., and B. A. Manja, Shear layers above the break in bottom topography, Geophys. Astrophys. Fluid Dyn., 14, 45-60, 1979.

Johnson, J. A., and B. A. Manja, Longshore currents over the shelf break, Rapp. P. V. Reun. Cons. Int. Explor. Mer, 180, 73-74, 1982.

Johnson, J. A., and A. J. G. Nurser, A model of secondary upwelling over the shelf break, Geophys. Astrophys. Fluid Dyn., 23, 301-320, 1983.

Johnson, J. A., and A. J. G. Nurser, A model of secondary upwelling over the shelf break, II, Geophys. Astrophys. Fluid Dyn., 28, 161-170, 1984.

Jones, P. G. W., The variability of oceanographic observations off the coast of north-west Africa, Deep Sea Res., 19, 405-431, 1972.

Jones, P. G. W., and A. R. Folkard, Chemical oceanographic observations off the coast of northwest Africa with special reference to the process of upwelling, Rapp. P. V. Reun. Cons. Int. Explor. Mer., 159, 38-60, 1970.

Kinder, T. H., and L. K. Coachman, The front overlaying the continental slope in the east Bering Sea, J. Geophys. Res., 83(C9), 4551-4559, 1978.

Lagerloef, G. S., R. D. Muench, and J. D. Schumacher, Low-frequency variations in currents near the shelf-break: Northeast Gulf of Alaska, J. Phys. Oceanogr., 11, 627-638, 1981.

Leming, T. D., and C. N. K. Mooers, Cold water intrusions and upwelling near Cape Canaveral, Florida, in Coastal Upwelling, Coastal Estuarine Sci. Ser., vol. 1, edited by F. A. Richards, pp. 63-71, AGU, Washington, D.C., 1981.

Lill, C. C., Upwelling over the shelf break, J. Phys. Oceanogr., 9, 1044-1047, 1979.

Meincke, J., E. Mittelstaedt, K. Huber, and K. P. Koltermann, Currents and stratification in the upwelling area off northwest Africa, Meereskdl. Beob. Ergeb. Dtsch. Hydrogr. Inst., 44, 117 pp., 1975.

Mittelstaedt, E., The upwelling area off northwest Africa--A description of phenomena related to coastal upwelling, Prog. Oceanogr., 12, 307-332, 1983.

Mittelstaedt, E., and I, Hamann, The coastal circulation off Mauritania, Dtsch. Hydrogr. Z., 34, 81-118, 1981.

Mittelstaedt, E., D. Pillsbury, and R. L. Smith, Flow patterns in the Northwest African upwelling area, Dtsch. Hydrogr. Z., 28, 145-167, 1975.

O'Brien, J. J., and G. W. Heburn, The state-of-the-art in coastal ocean modelling: A numerical model of coastal upwelling off Peru--including mixed layer dynamics, in Coastal Oceanography, edited by H. Gade, A. Edwards, and H. Svendsen, pp. 113-164, Plenum, New York, 1983.

O'Brien, J. J., and H. E. Hurlburt, A numerical model of coastal upwelling, J. Phys. Oceanogr., 2, 14-26, 1972.

Petrie, B. D., Current response at the shelf break to transient wind forcing, J. Geophys. Res., 88(C14), 9567-9578, 1983.

Pingree, R. D., and G. T. Mardell, Slope turbulence, internal waves and phyto plankton growth at the Celtic Sea shelf-break, Philos. Trans. R. Soc. London, Ser. A, 302, 663-682, 1981.

Preller, R., and J. J. O'Brien, The influence of bottom topography on upwelling off Peru, J. Phys. Oceanogr., 10, 1377-1398, 1980.

Schumacher, J. D., and T. H. Kinder, Low-frequency current regimes over the Bering Sea shelf, J. Phys. Oceanogr., 13, 607-623, 1983.

Shaw, P.-T., and G. T. Csanady, Self-advection of density perturbations on a sloping continental shelf, J. Phys. Oceanogr., 13, 769-782, 1983.

Smith, R. L., A comparison of the structure and variability of the flow field in Three Coastal upwelling regions: Oregon, Northwest Africa and Peru, in Coastal Upwelling, Coastal Estuarine Sci. Ser., vol. 1, edited by F. A. Richards, pp. 107-118, AGU, Washington, D.C., 1981.

Suginohara, N., Coastal upwelling: Onshore-offshore circulation, equatorward coastal jet and poleward undercurrent over a continental shelf-slope, J. Phys. Oceanogr., 12, 272-284, 1982.

Thompson, J. D., and J. J. O'Brien, Time-dependent coastal upwelling, J. Phys. Oceanogr., 3, 33-46, 1973.

Thompson, R. O. R. Y., and G. Veronis, Poleward

boundary current off Western Australia, Aust. J. Mar. Freshwater Res., 34, 173-185, 1983.

Tomczak, M., and R. H. Käse, A linear theory of stationary coastal upwelling in a continuously stratified ocean with an unstratified shelf area, J. Mar. Res., 32, 365-376, 1974.

Van Foreest, D., and G. B. Brundrit, A two-mode numerical model with applications to coastal upwelling, Prog. Oceanogr., 11, 329-392, 1982.

Whitehead, J. A., Laboratory models of circulation in shallow seas, Philos. Trans. R. Soc. London, Ser. A, 302, 583-595, 1981.

# COASTAL AND ESTUARINE FRONTS

J. H. Simpson

University College of North Wales, Marine Science Laboratories, Menai Bridge
Gwynedd, LL59 5EY, United Kingdom

I. D. James

Institute of Oceanographic Sciences, Bidston Observatory
Birkenhead L43 7RA, England

Abstract. Fronts are regions of intensified
horizontal gradients which occur widely in shelf
seas and estuaries. They constitute important
features of the structure and circulation and may
play a critical role in relation to biological
production as well as being of practical importance in problems of waste disposal in coastal
seas. The basic processes responsible for the
generation of several types of fronts have been
identified, and useful generalizations have been
achieved, particularly about the location and
duration of the shelf sea fronts produced by tidal
stirring. We also have an increasing understanding of the influence of freshwater runoff in producing a variety of fronts in estuaries and at the
boundaries of coastal currents. Knowledge of the
internal dynamics of fronts, however, remains
limited, and we describe the various theoretical
models which have been developed to explore the
principal dynamical processes operating in
fronts. Two-dimensional diagnostic models provide
predictions of the cross-frontal vertical plane
circulation, but these may be of only limited
relevance because baroclinic instability processes
produce large-scale horizontal eddies which dominate the horizontal transport. Modeling effort is
therefore increasingly being concentrated on
three-dimensional density-advecting models which
may be used to investigate the details of instability motions. The validation of such models will
require improved near-synoptic measurements of
frontal regions using both satellite infrared
observations and undulating conductivity-tempera-
ture-depth measurements.

## 1. Introduction
### Theoretical and Practical Significance of Fronts

An extended region of intensified gradients, in
which ocean properties change more rapidly with
horizontal distance than in the surrounding waters, is commonly known as a "front" or a "frontal

zone." The terms are sufficiently imprecise to
allow their use to describe almost all horizontal
boundaries of proven or suspected significance in
the ocean. Within this wide range of phenomena, a
number of well-defined classes of ocean fronts are
now recognized, and their importance as structural
features of the ocean is under active investigation.

Frontal zones are usually distinguished from
the rest of the ocean by the relatively large
velocities and velocity gradients which exist
there. A feature of most frontal systems is a
marked convergence in the horizontal flow usually
at the surface. Associated with this convergence
is a pattern of pronounced vertical motions making
the front an area of enhanced vertical transfer of
momentum and other properties. Horizontal velocities, too, tend to be increased, particularly in
large-scale fronts where the geostrophic constraint requires an along-front flow concentrated
in the region of horizontal density change.

Fronts, especially those which are consistent
in position, are often thought of as dividing the
ocean into different regimes with separate water
masses. This should not, however, be taken to
imply that fronts are necessarily a barrier to
horizontal exchange. They may, on the contrary,
experience strong cross-frontal fluxes of properties through the operation of horizontal mixing
processes on the large gradients that obtain
there.

Cross-frontal mixing is one candidate mechanism
for explaining anomalies in phytoplankton and
other biomass distributions at fronts. The velocity field, particularly the vertical plane circulation, may also be important in this context, in
bringing about accumulations of buoyant or mobile
organisms in the frontal zone.

Such influences on the distribution of biomass,
and possibly on its production, have clear implications for fisheries biology. In addition,
fronts may also have practical significance in

63

other aspects of the management of the shelf seas. Their presence will modify the circulation and mixing processes in ways that may have significant influence on the dispersal of pollutants discharged into coastal waters. The proximity of a frontal zone may therefore be an important consideration in, for example, the siting of a nuclear power station or an offshore dumping operation. For buoyant pollutants (e.g., hydrocarbons) the convergence associated with the front will certainly inhibit dispersion and lead to a concentration of pollutants along the line of the front. Such accumulations, which frequently indicate the presence of a front, may even prove useful, as preferred sampling sites, in the detection of incipient pollution problems.

Apart from these practical considerations, fronts invite study by oceanographers because their existence and structure pose a number of interesting and fundamental questions. Among these is the basic puzzle of why the ocean shows such a strong tendency to organize much of its variability into regions of concentrated gradients which may persist over long periods of time. This tendency is manifest on a wide range of scales and is somewhat akin to the concentration of gradients in the vertical fine structure of the ocean. The internal structure of the fronts and the dynamics of cross-frontal circulation and eddy instabilities also present a number of challenging questions to theoreticians and experimental oceanographers alike.

To many of these questions there are, as yet, only partial answers. In this paper we shall first endeavor to summarize what is known from observations of those fronts which occur in shallow seas and estuaries. We shall consider both the characteristic thermal fronts of the shelf seas, in which temperature differences are responsible for the density structue, and haline fronts, occurring in and close to estuaries, where freshwater outflow is the dominant influence. Then, through a discussion of recent models of frontal behavior, we shall try to illustrate some of the dynamical and mixing processes operating within fronts.

## 2.  Problems in Observing Frontal Phenomena

The observational study of fronts has until recently been impeded by the slow rate at which data can be acquired in traditional oceanographic survey work. The full delineation of a frontal zone and its internal structure involves the measurement of at least two properties over a three-dimensional spatial grid. Frequently, the time taken for such surveys is so long as to allow significant evolution of the front, so that the results cannot be treated as a synoptic picture of the front. This deficiency may be apparent even when observing frontal features of limited scale in estuaries, but it becomes most pronounced when we are dealing with relatively large scale (~100 km long) fronts in the shelf seas. The

basic problems are the slow speed of survey vessels (~10 knots) and their high cost, which often means that only one ship can be utilized in a frontal survey. These factors, together with the slow procedure of oceanographic station work, have severely restricted the extent and usefulness of data sets on fronts. They have also prompted workers in the field to take advantage of two recent technical developments which have helped to circumvent the above constraints.

The use of radiation thermometers in aircraft represented a marked improvement over ship surveys of sea surface temperature (SST). Temperature resolution of ~ 0.1°C can be achieved in rapid surveys which yield results which are almost synoptic. Using calibration data from surface ships in the survey area, thermal maps with an accuracy of a few tenths of a degree Celsius can be achieved.

Combined ship-aircraft studies of SST distributions have proved valuable in identifying the extent and configuration of frontal zones, but an even more useful application of infrared (IR) methods has come with the incorporation of scanning radiometers in polar-orbiting satellites. The NOAA series of satellites (from NOAA 4 on) have been equipped with IR scanners in which the compromise between temperature resolution (~0.2°C) and spatial resolution (~1 km) is near ideal for the observation of fronts in the shelf seas. Using such satellites, the synoptic monitoring of frontal zones at frequent intervals may be maintained over long periods of time without great increase in costs.

The method is, unfortunately, subject to interference by cloud cover. It is also important to remember that because of absorption and radiation by water vapor in the atmosphere, the observed radiation temperatures do not correspond to absolute SST. The observed radiation, moreover, comes from the top 1 mm of the ocean, in which the horizontal distribution of temperature may not be characteristic of the surface layer.

A second innovation, that of the undulating conductivity, temperature, and depth (CTD) system, has helped to improve our knowledge of the subsurface structure of temperature and salinity within the frontal zone. A conventional set of CTD sensors is contained in a vehicle which can be made to undulate behind the ship as it proceeds at full speed. Profile data down to depths of more than 100 m can be obtained at regular intervals along the track, giving an effective horizontal resolution of a few hundred meters. Using such devices, it has been possible to obtain vertical plane temperature and salinity sections through frontal regions of much higher definition than is possible with conventional station procedures. Adding a fluorimeter to the sensor package permits estimation of the distribution of chlorophyll and associated pigments in the frontal zone. Such parallel biological and physical data offer great promise in the future assessment of the role of fronts in the primary production of the shelf seas.

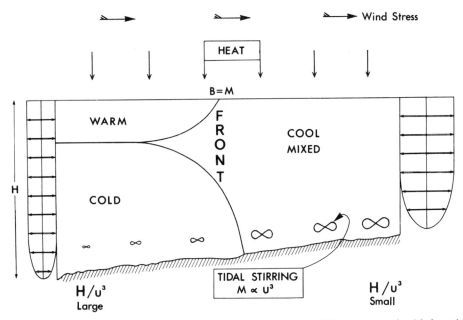

Fig. 1. Schematic of the competition between seasonal heating and tidal stirring.

### 3. Fronts Produced by Seasonal Heating and Tidal Stirring

#### 3.1. The Basic Mechanism

Recent studies of the shelf seas have revealed the important role played by tidal stirring in promoting vertical mixing in these areas. Most of the tidal energy which enters shelf seas from the deep ocean is ultimately dissipated as heat through the action of strong frictional stresses which result from vigorous tidal streams flowing over the seabed. During periods of surface heating, however, a small proportion of the energy lost from the tidal flow may be used to work against the buoyancy forces and drive the heat downward.

In winter, heat loss at the sea surface causes a convectively mixed regime in which the temperature is almost vertically uniform throughout most of the shelf seas. Heat input becomes positive soon after the vernal equinox and is largely concentrated in the surface layers. According to Ivanov [1977], in coastal waters ~90% of the total energy input occurs in the top 5 m of the water column. Tidal stirring, together with a contribution from the wind stress on the surface, will act to redistribute this surface input and oppose the development of a warm, low-density surface layer. Over much of the shelf the stirring is sufficient to maintain the vertical homogeneity of the water column. Elsewhere, as shown schematically in Figure 1, lower levels of mixing permit the establishment of stratification, which soon becomes sufficiently robust to enable it to persist throughout the summer season. The result is that the shelf is partitioned into mixed and stratified regimes and these are separated by rather narrow frontal zones.

For the case where tidal stirring predominates over wind mixing, the balance between stirring and heating is represented by the parameter $Qh/U_s^3$ = const along a front, where $Q$ is the heating rate, $h$ is the water depth, and $U_s$ is the surface tidal stream amplitude at springs.

This criterion for the location of fronts [Simpson and Hunter, 1974] follows from basic energy considerations discussed in section 7.1. Its validity has been tested by comparison with both ship and remote sensing data.

For a limited region of the shelf, $Q$ is often a function of time only, and so the quantity $h/U_s^3$ should determine frontal positions. $U_s^3$ was originally based on the surface tidal stream amplitude at springs because this quantity was available in tidal atlases. It is now, however, usually preferable to use the closely related parameter

$$\chi = h/\overline{|\underset{\sim}{U}|^3}$$

(units of $m^{-2}s^3$), where $\underset{\sim}{U}$ is the depth mean tidal velocity vector and the overbar denotes a mean over the tidal cycle. The distribution of $\chi$ is conveniently determined for numerical two-dimensional models of the tidal flow [e.g., Pingree and Griffiths, 1978].

An analysis of all available temperature and salinity data for large areas of the shelf around the United Kingdom in terms of the potential energy anomaly [Simpson et al., 1977] shows the close relation of stratification to the $\chi$ distribution, with fronts occurring at a critical value of $\log_{10} \chi \simeq 2.7$. This result is supported by evi-

Fig. 2.  Satellite infrared image of United Kingdom shelf area from NOAA 6, May 16, 1980, 1901 UT.

dence from satellite IR imagery, in which the fronts appear as abrupt changes in SST. Figure 2 shows a particularly clear image of the United Kingdom area, while Figure 3 shows composites of frontal positions from sequences of images compared with the $h/U^3$ distribution.

The predominant influence of this parameter is further emphasized in Figure 4, which is a plot of the potential energy anomaly

$$\phi = \frac{1}{h}\int_{-h}^{0}(\bar{\rho} - \rho)gz\ dz \qquad \bar{\rho} = \frac{1}{h}\int_{-h}^{0}\rho\ dz$$

determined from station data versus $h/U^3$.

Results from other tidally dominated areas are similar. Garrett et al. [1978], for example, have shown that the distribution of stratification in the Gulf of Maine and the Bay of Fundy is closely related to the $h/U^3$ distribution derived from a numerical model of the tidal flow. The critical value for transition to vertical mixing is almost identical to that observed for the European shelf.

## 3.2. Frontal Movements

The fronts are not, of course, rigidly tied to $h/U^3$ contours. They exhibit some variability of position, part of which may be attributed to simple tidal advection. When this has been removed, the remaining movement is less easy to understand. Only a small contribution to the variance of frontal positions is apparently made by the springs to neaps cycle in the tidal range, in spite of the fact that there are changes in U by a factor of ~2 for most of the shelf seas.

A continuous balance between heating and stirring at the front would require it to move between equilibrium positions at which the average mixing power differed by a factor of ~$(2)^3 = 8$. However, an equilibrium balance will not be attained at the springs phase of the cycle because of stored buoyancy accumulated during the neaps period. As shown schematically in Figure 5, the condition for the springs position is that the average rate of stirring is sufficient to balance the rate of buoyancy input by heating. This condition implies that the ratio of mixing powers at the neaps and springs positions should be

$$\frac{P_n}{P_s} = \frac{(2 + 3\ e^2)}{2(1 - e)^3} \qquad (1)$$

where e is the fractional change in U between the average tidal range and springs [Simpson and Bowers, 1981]. For a representative value of e = 0.3, we have $P_n/P_s$ = 3.31, which is just over half the adjustment required for an equilibrium balance.

The observed movement with the springs-neaps cycle is even smaller. Figure 6a summarizes the results of observations of the positions of fronts in the Irish Sea along reference lines approximately perpendicular to the fronts. Average fron-

tal positions, in the form of the ratio of $h/u^3$ at the frontal position to its mean value, are plotted against the tidal range factor F. The F values are those occurring 2 days before the satellite observations, a time lag which gives the maximum correlation. What movement there is occurs mostly at neaps, and the total displacement is only ~40% of that predicted by equation (1). This restricted response to the springs-neaps stirring cycle suggests the operation of a significant degree of feedback in the stratification process, so that the efficiency of mixing is reduced by increasing stability. One-dimensional mixing models incorporating such a feedback process give an improved account of the observed movement (see section 7.1 and also Simpson and Bowers [1981] and Simpson [1981]). Such models may also offer an explanation for the observed seasonal behavior of frontal positions. Figure 6b shows the mean frontal positions from IR observations versus day number together with the predicted cycle based on a variable efficiency model. Although the heating rate Q becomes positive soon after the vernal equinox (about day 90), the surface gradients do not generally appear strong enough to be identified in the IR imagery until mid-April (day 110) at the earliest. By that time, the fronts have already advanced to a position close to that in which they remain until the autumnal equinox. Feedback serves both to accelerate the advance of the front in the spring and to delay the breakdown of stratification at the end of the season.

The fact that the mean positions of fronts are so consistently in accord with the $h/u^3$ contours has implications for the residual velocity field. The $h/u^3$ criterion is derived from energy arguments for a one-dimensional model of local vertical exchange in which it is assumed that there is no net horizontal flow. The success of the model therefore depends on the residual velocity component normal to fronts being less than a value of ~1 cm/s defined by the ratio of the advective terms to the heating input [Simpson, 1981]. This puts an important constraint on the residual flow field in many areas and suggests that deviations of mean frontal positions from the appropriate $h/u^3$ contour may provide evidence of sustained residual currents.

While the mean positions of fronts averaged over intervals of F (Figure 6a) or time (Figure 6b) are remarkably consistent, there is still significant variability in the individual observations of position. The variability, which corresponds to an rms displacement of ~7 km, is apparently random in character and may be a consequence of the eddy instabilities, which are apparent in many IR images and with which we shall deal in the next section.

## 3.3. The Structure of Fronts

In assessing our present picture of the internal structure of frontal zones it is important to

Fig. 3a. Composite of all cloud-free images during May 1978 showing observed frontal positions uncorrected for tidal displacement.

Fig. 3b. Same as Figure 3a but for the period May–September 1980.

recall the circumstances of the measurements. We are attempting to determine the temperature, salinity, velocity, and other distributions in a volume which is continuously moving with the mean tidal advection and may be significantly distorted by horizontal and vertical shears in the tidal currents. In many frontal surveys, the relatively small horizontal scales are not fully resolved by the sampling scheme, and the slow rate of sampling

means that significant evolution of the structure may have occurred during the survey.

In the case of the temperature and salinity fields, the above mentioned undulating CTD has provided the most comprehensive sections of frontal regions to date, although the complexity and cost of the technique, and the large amount of data reduction required, has meant that only a limited number of surveys have been made in this way.

Figure 7 represents the temperature structure of the western Irish Sea front as observed with a

Fig. 3c. Frontal positions predicted by h/u³
parameter. The heavy lines represent the contour
$\log_{10}$ h/$|\underset{\sim}{U}_2|^3 \simeq 2.7$, where $\underset{\sim}{U}_2$ is the depth mean
tidal velocity. Lighter lines represent contours
at levels of 2.7 ± 0.05 for the more prominent
fronts (adapted from Pingree and Griffiths [1978]
and Proctor [1981]).

(~ 0.5°C). By contrast, the situation in the west
is one of intense stratification, with temperature
differences between the surface layer and the
water at 50 m depth of >6°C. This temperature
stratification dominates over salinity in controlling the density in this frontal region as in many
others, but there are cases where salinity structure also makes a major contribution (e.g., the
Islay Front [see Simpson et al., 1979]).

In the frontal region itself the pycnocline is
tilted up to the surface, where it outcrops in a
region where the maximum horizontal gradient exceeds 1.5°C/km. Close to the gradient maximum, on
the mixed side of the front, there is a minimum in
surface temperature, and associated with this we
see an upward displacement of the 10°, 10.5°, and
11°C isotherms. This minimum, though small in
magnitude (~0.1°C below the mixed water surface
temperature) is common to many such sections and
is frequently observed in data from continuous
surface temperature recorders. Such effects are
strongly suggestive of the upwelling of cold water
from the lower layer on the stratified side. In
the case of the western Irish Sea front, the water
in the surface minimum is often the coldest surface water in the Irish Sea and so can only be
sustained by vertical transfer.

This indication of upwelling is one of the few
observational clues that we have to the form of

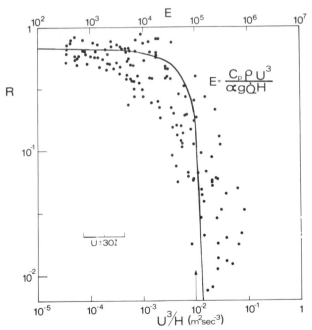

Fig. 4. Plot of the potential energy anomaly in
nondimensional form versus u³/h. R is defined by
R = $\phi/\phi_m$ , where $\phi_m$ is the value of $\phi$ which
would occur in the absence of stirring. The data
are from a survey of the Irish Sea and Celtic Seas
during June 1973. The curve represents a model
based on tidal stirring with a constant efficiency
of mixing.

Brown CTD mounted in a modified Batfish vehicle
[Dessureault, 1976]. At the eastern limits of
this section the vertical mixing is seen to be
almost complete, with temperature differences of
only a few hundredths of a degree Celsius over
most of the water column. Only near the surface
is there any significant temperature change

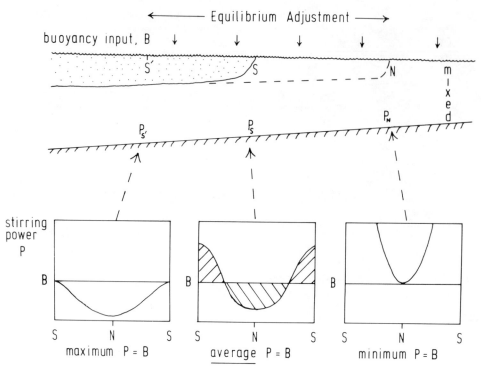

Fig. 5. Schematic of the relative magnitudes of the buoyancy input B and stirring power P over the fortnightly cycle as the tidal streams vary according to $U_2 = U_0 (1 + e \cos \sigma t)$; $\sigma = 2\pi/14.5$ days. Note that the "springs" position is controlled by the average rather than the maximum stirring rate and that the maximum advance of the front into the stratified region will occur about 2–3 days after springs.

vertical plane circulation associated with the fronts. A second is provided by the existence of a marked convergence zone close to the region of maximum gradient. Accumulations of surface material, sometimes including concentrations of organisms such as jellyfish, are frequently observed in crossing fronts. In calm weather this labeling by flotsam may be so marked as to make the front clearly visible from an overflying aircraft.

Such evidence leads to the postulation of a vertical plane circulation pattern of the form shown in Figure 8. The required convergence in the high-gradient region with a divergence and upwelling in the mixed water may be the result of frictional influence on the along-front geostrophic flow. The latter is required to balance the pressure gradients associated with the density gradients in the front. Frictional forces will modify the geostrophic velocity field, leaving an unbalanced component of pressure gradient normal to the front which results in the vertical plane circulation. This mechanism is clearly illustrated in the two-dimensional models of James [1978] and Garrett and Loder [1981] (see section 7.2). It must be remembered, however, that there is no direct evidence for the vertical plane flow, measurements of which would present formidable difficulties. There are, at the same time, considerable doubts about the applicability of two-dimen-

sional models of frontal structure because of the important role apparently played by large-scale eddies which develop from instabilities in the along-front flow. Observations using radio-tracked drogues, aimed at measuring the velocity of the along-front jet, revealed a complex horizontal velocity field which could not generally be rationalized in terms of quasi-rectilinear geostrophic flow [Simpson et al., 1978]. In only one case, that of the Islay front [Simpson et al., 1979], was a sustained flow resembling the frontal jet observed, although even in this case a clear geostrophic balance could not be established.

The complex nature of the flow generally observed results from the development of large-scale disturbances due to baroclinic and possibly other instabilities. Indications of such processes can be seen in a number of satellite IR images which show "eddy" structures with horizontal scales of 20–40 km and a time scale for development of ~3 days. The precise nature of these motions is not always clear from the limited sequences available, and different interpretations may be given in terms of simple cyclonic eddies [e.g., Pingree, 1978] or vortex pairs [James, 1981] which are analogous to the structures observed by Griffiths and Linden [1981a] in tank experiments on baroclinic instability.

Another basic question, still to be answered,

70    SIMPSON AND JAMES

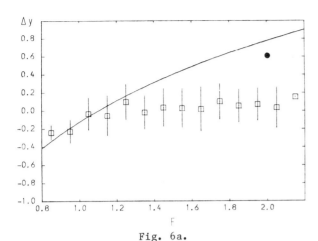

Fig. 6a.

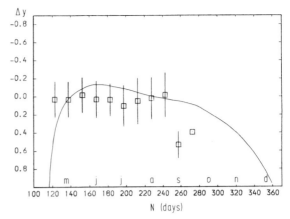

Fig. 6b.

Fig. 6. Variations in frontal positions. (a) Plot of $\Delta y$ versus F for 430 observations of the Islay, western Irish Sea, and Celtic Sea fronts in the months May to August during the years 1976-1980. $\Delta y = y - \bar{y}$, where y is the value of $\log_{10} \chi$ at the observed frontal position and $\bar{y}$ the average of all points for the front concerned. F is the tidal range factor, which is the tidal range at Liverpool normalized so that the mean neaps value is 1.0. The values of F are those occurring 2 days prior to the observation of frontal position, a time lag which yields the maximum correlation. The large number of points has been reduced by plotting the average $\Delta y$ in each interval of 0.1 in F and showing the variability as a vertical line of length $\pm 1$ standard deviation. The observed movement occurs mainly near neaps and is much less than that required for the equilibrium adjustment shown by the solid curve. The circle represents the limit of adjustment when allowance is made for stored buoyancy. (b) Plot of $\Delta y$ for the same data set versus Julian day number. The squares represent the values of $\Delta y$ averaged over 15-day intervals, and the vertical lines show the variability within each interval as $\pm 1$ standard deviation. The solid curve shows the frontal position predicted by a variable efficiency model [Simpson and Bowers, 1981].

is whether the eddies are confined to frontal regions or are simply manifest there because of the temperature contrast. In the tank experiments of Griffiths and Linden [1981a], eddies generated at a frontal boundary were observed to migrate into the interior stratified region, creating a steady random distribution of eddy motions.

## 3.4. Biological Implications

The recognition of clearly defined mixed and stratified regimes, separated by frontal zones, has provided the basis for a new assessment of biological processes in the shelf seas. From the point of view of a marine organism, the mixed and stratified areas represent distinctly different environments, with the fronts forming the almost fixed geographical "fences" between them.

The environments differ primarily in the level of vertical mixing. In the mixed regime, where there is negligible vertical stability, a passive phytoplankton organism will be carried up and down throughout the water column by the turbulent velocity field. The available light will be intermittent, and its average value may be low if the nondimensional depth kh is large (where k is the diffuse attenuation coefficient for natural radiation). The strong vertical mixing itself can

contribute to a large value of k by maintaining suspensions of fine inorganic particles. Vertical mixing also provides a diffusive flux of nutrients from recycling processes occurring in the bottom boundary region. Taken together with the constraints on the radiation supply, this implies that the growth of phytoplankton is generally not limited by nutrient availability in the mixed waters.

In the stratified water, however, the onset of stability in the spring is followed by a rapid exploitation of the available nutrients in the surface layer, and with the pycnocline forming an effective barrier to vertical diffusion, depletion follows. At the same time, the radiation supply may be improved by the settling out of inorganic particles which follows the onset of stability. Extreme contrasts in seston concentrations are frequently observed in April and May at the western Irish Sea front. Figure 9a shows a coastal zone color scanner (CZCS) image from April 1980 when the front has just become established. Surface temperature contrast at this time is $\sim 0.5°C$, so the front is scarcely perceptible in the IR imagery. In all the narrow-band visible channels of CZCS, however, it is clearly visible, and it can also be seen in broadband imagery from the NOAA 6 AVHRR (Figure 9b).

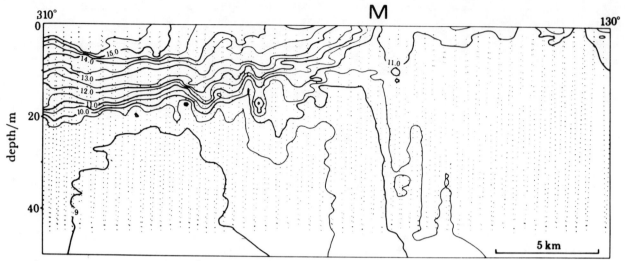

Fig. 7. Vertical section of the temperature structure in the western Irish Sea front (A in Figure 2) on July 7, 1976. The dots represent data points used in the analysis. Data point separation is ~1 m in the vertical and 250-500 m in the horizontal. M is the surface temperature minimum. The salinity variation over the section, determined in parallel with the temperature measurements, was small (34.15 ± 0.1), so that the density field closely resembles the temperature section.

These contrasting differences in radiation climate and nutrient supply have led to a number of hypotheses about the possible importance of the frontal region as one in which a better balance of light and nutrient availability would permit greater production [Pingree et al., 1975] than is possible on either side of the frontal zone.

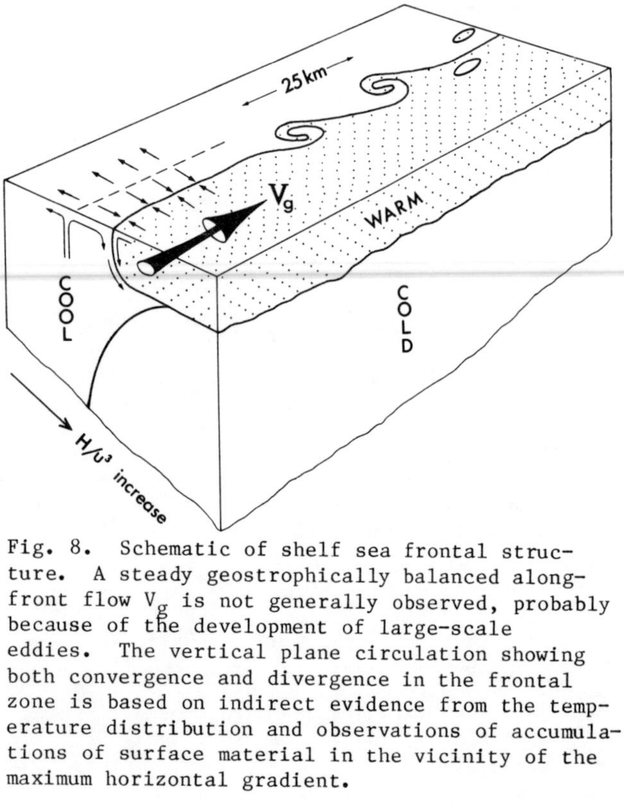

Approaching the front from the stratified side, we might expect to see an increasing upward flux of nutrients through the pycnocline as stability decreases. This flux should promote production in the surface layer, with a maximum occurring just before the final breakdown of stratification. Another possibility is that the periodic variations in frontal position during the springs-neaps cycle will result in a nutrient-rich body of water being held close to the surface following the advance of the front, thus creating favorable conditions for growth. Nutrient-rich water from the mixed side may also intrude along the pycnocline and become stabilized at a level where the light intensity will allow full exploitation of this input. We shall discuss this possiblity further in relation to mixing around islands.

It has also been suggested that the mixing of the surface waters from the two sides of the front may lead to the creation of a mixture which provides all the necessary nutrients and "micronutrients" for growth, one or more of which may be missing from the separate water masses. Attempts to test this complementation hypothesis have been described by Savidge [1976].

Once an increase in primary production commences at a front, there may be biological feedback mechanisms that further accelerate growth. Floodgate et al. [1981] have proposed that an initial upsurge in production will attract herbivores and eventually higher trophic levels into the frontal zone with a consequent increase in the recycling rate for nutrients.

Fig. 8. Schematic of shelf sea frontal structure. A steady geostrophically balanced along-front flow $V_g$ is not generally observed, probably because of the development of large-scale eddies. The vertical plane circulation showing both convergence and divergence in the frontal zone is based on indirect evidence from the temperature distribution and observations of accumulations of surface material in the vicinity of the maximum horizontal gradient.

The testing of the basic notion that fronts are generally regions of augmented production is made difficult by the short time scales and high variability of biological processes. A prima facie case can be made in support of the idea that fronts tend to be regions of high standing crop of phytoplankton, especially those situated in water deeper than 50 m. [Holligan, 1981]. The observed standing crop is, however, determined both by primary productivity and by grazing activity by herbivores, and the latter is largely unknown. It is not therefore possible to relate production to standing crop, and direct measurements of photosynthesis rates have been obtained in only a limited number of cases.

Extreme maxima in standing crop, while well represented in the literature, may be very unrepresentative of average levels at fronts, and there is a need for systematic repeated sampling in stratified, mixed, and frontal regimes to assess average levels of phytoplankton density and production rates. Even if average levels of standing crop are not very much higher than in the mixed and stratified waters, as suggested by recent results from the western Irish Sea front [Mitchelson et al., 1985], higher rates of production would seem to be necessary to supply the demands of higher organisms whose concentration at fronts is clearly signaled by intense fishing activity.

A frontal zone which may prove of particular interest from the biological point of view is that which surrounds an island in a stratified sea. The presence of the island causes an acceleration of the tidal flow, which leads to a local increase in vertical mixing. In cases where the primary tidal flow is strong enough, this results in the establishment of vertically mixed regions which are bounded by frontal zones [Simpson, 1981].

Figure 10a shows regions of cold water produced in this way by the presence of the Scilly Isles, in the Celtic Sea. The distribution of SST bears a close relation to the local contours of $h/u^3$, obtained from a fine-scale numerical model (Figure 10b). At times this basic mixing pattern is distorted by the effect of the mean flow, which tends to advect the mixed water away from the island as is evident from some IR images [Simpson, 1981].

The mixture of waters produced by the island's stirring action is relatively high in nutrients because of the recruitment of water from the subthermocline region where levels are undepleted. With a density intermediate between the upper and lower layers, the mixed water will tend to spread into the thermocline producing favorable conditions for phytoplankton growth. Observational evidence for such a mechanism was obtained in a detailed survey of the Scilly Isles region during July 1979 [Simpson et al., 1982]. The concentration of phytoplankton biomass was generally greater close to the islands than in the surrounding stratified water, but there was a particular intensification in the pycnocline around the islands where chlorophyll levels up to 30 µg/L were re-

corded. Estimates of primary productivity were also much higher than in the Celtic Sea away from the islands, the total enhancement of production being equivalent to that of an area, unaffected by island stirring, roughly 150 times the area of the Scilly Isles.

Similar effects from tidal stirring by other islands and around headlands may be anticipated and could make a major contribution to the overall production of the shelf seas.

## 4. Fronts Produced by Freshwater Inputs

In our consideration of the shelf sea fronts of the last section, it was assumed that the main source of buoyancy is heating through the sea surface. In most areas, inputs of buoyancy through precipitation are, on average, unimportant in relation to the heat flux, but local injections of buoyancy in the form of river runoff may reverse this situation and result in pronounced haline stratification. Associated with this stratification, there commonly exist frontal features with strong horizontal salinity gradients, which, depending on the magnitude of the river discharge and the intensity of tidal flow, range from large features extending far into coastal waters down to small-scale fronts located within the estuary itself.

### 4.1. Plume Fronts

Where tidal flows are weak, low-salinity water leaving the estuary will spread seaward as a thin surface layer in response to the buoyancy force. The resulting plume will be bounded on both sides by frontal regions as observed, for example, in the Mississippi discharge by Wright and Coleman [1971] or, on a much smaller scale, in a Norwegian fjord by McClimans [1978]. Perhaps the best documented example of such a plume front is that reported by Garvine and Monk [1974] (see also Garvine [1979a, b]). At times of high discharge the Connecticut River flows into Long Island Sound on the ebb tide as a buoyant plume ~2 m thick. The offshore boundary of the plume is characterized by extreme salinity gradients with obvious color changes and accumulations of surface material. Strong convergent velocities (20-50 cm/s) are found at the surface on both sides of the front, with vigorous sinking motions at the front. This vertical plane motion induces a downward entrainment of fluid from the surface brackish layer, leading to vertical mixing and the eradication of the frontal structure over a period of several hours.

An extreme case of plume development is that associated with the Amazon, whose outflow is so great as to permanently exclude salt water from its estuary. The brackish water, produced by tidal mixing in shallow water near the river mouth, spreads as a plume and is at the same time advected along the coast of the Guianas, forming an extensive and complex system of fronts. At

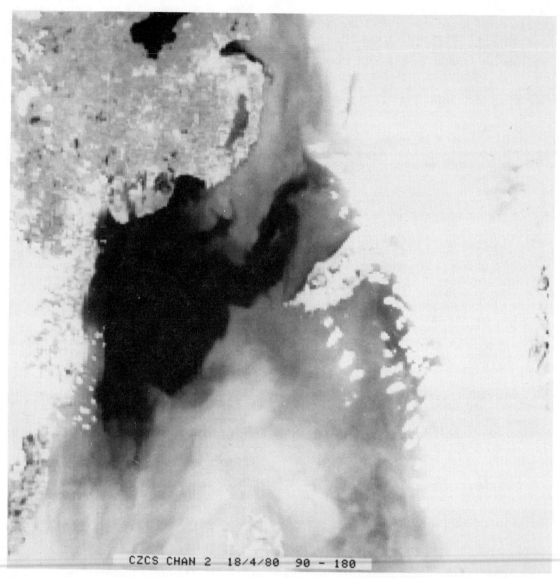

CZCS CHAN 2  18/4/80  90 - 180

Fig. 9a

Fig. 9. Visible band images of the western Irish Sea frontal region: (a) Coastal zone color channel 2 image on April 18, 1980, 1038 UT ($\lambda = 520 \pm 10$ nm). (b) TIROS N AVHRR channel 1 on April 4, 1980, 1456 UT (broadband visible sensor). Dark areas to the northwest are recently stratified and have low concentrations of scattering particles in contrast to mixed areas (whiter) to the southeast. The high concentration gradient coincides with the breakdown of stratification at the front.

times, patches of brackish water become detached from the outflow and form isolated lenses [Gibbs, 1970].

Away from the equator, large-scale buoyant outflows are subject to the Coriolis force. Where mixing processes are weak and the flow is essentially frictionless, a coastal current in geostrophic balance is observed. The northward flowing Norwegian current, driven by buoyancy inputs from the Baltic and the Norwegian fjords, is an example

of such a flow. The outer edge of this current is frequently observed to take the form of a sharp front with large salinity gradients and associated temperature gradients which may be observed in satellite IR imagery.

As in the case of the tidal stirring fronts, the geostrophic flow is apparently unstable and commonly exhibits large meanders [Mork, 1981]. These frontal waves have been observed in sequences of satellite images which show them to

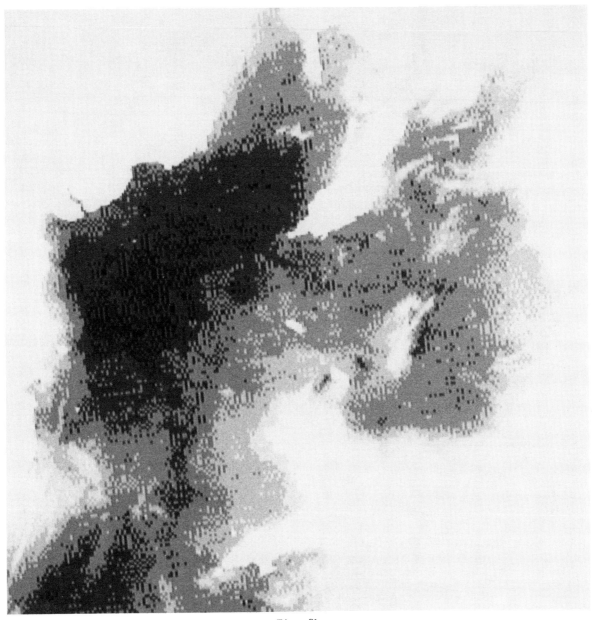

Fig. 9b

have a wavelength of ~80 km and propagation speeds of ~17 cm/s. Similar large-scale features are also observed in the East Greenland current, where the eddy structures are evident in Landsat imagery as convolutions of the ice margin [Wadhams et al., 1979].

In addition to buoyancy forcing, coastal currents may also be strongly influenced by wind stress, which in some circumstances may arrest the current. Aure and Saetre [1981] report such episodes in which the Norwegian coastal current is blocked by southwesterly winds and brackish water is forced to accumulate in the Skagerrak. When the opposing wind stress diminishes, the flow recommences in the form of an internal bore. The speed of propagation and width of the current correspond roughly to the phase speed and deformation radius of an internal Kelvin wave.

Although many examples of buoyancy-driven coastal currents have recently been reported, there have been few detailed studies of the frontal aspects of the outer edge of the current, so we do not yet know whether these are regions of particular biological significance. However, the

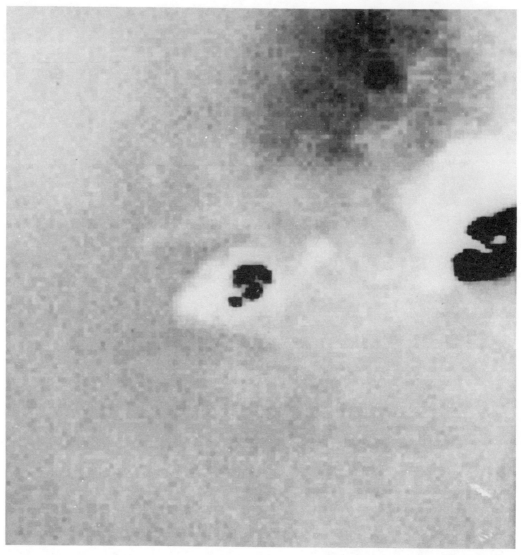

Fig. 10a

Fig. 10. (a) Infrared image of Scilly Isles region from TIROS N, June 6, 1979, 1455 UT.
(b) The distribution of $h/u^3$ for the Scilly Isles region based on the numerical model of
Argote [1982]. The contoured parameter is $\log_{10} h/|\underset{\sim}{U}_2|^3$, where $U_2$ is the $M_2$ tidal
stream vector determined from the model. The grid scale of the model is 2 km.

observation of high chlorophyll concentrations along the frontal boundary of the Norwegian coastal current [Mork, 1981] suggests the need to examine this possibility further.

Where tidal currents are larger, the fresh water will be more rapidly mixed with the ambient seawater, and the flow will be influenced by frictional forces. The distribution of stratification will then depend again on the availability of turbulent kinetic energy but in a more complex manner because the buoyancy is now input laterally from the river source rather than being uniformly distributed as in the case of surface heating. If the input of buoyancy due to river discharge F is

mixed down to depth h in an area A extending from the coast, then, in the absence of surface heat flux, the overall buoyancy-stirring balance may be expressed as

$$R = \frac{F\Delta\rho g}{2\epsilon kA} \frac{h}{(U^3)_a} = 1$$

where $\Delta\rho$ is the density contrast between river and seawater and $(U^3)_a$ is an average value over the area in which mixing occurs. The efficiency of mixing $\epsilon$ and drag coefficient k may be regarded as constants, so that again a form of $h/u^3$ appears as the controlling variable, although it should be

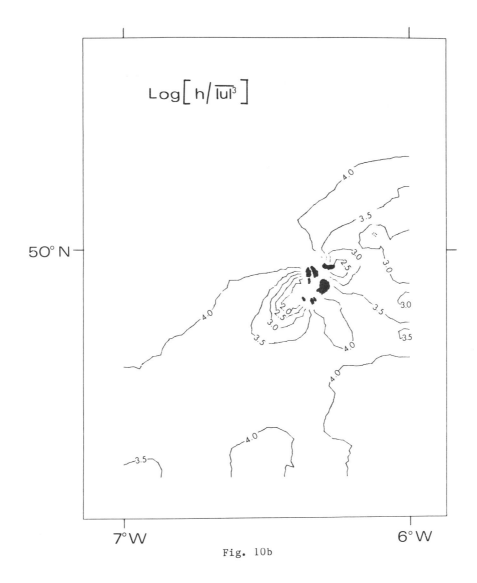

$$\text{Log}\left[\, h/\overline{|u|^3}\,\right]$$

Fig. 10b

noted that this condition is not strictly the same as that given in section 3.1. This type of balance, in which R is closely akin to Fischer's estuarine Richardson number, has been proposed by Bowman and Esaias [1981] to explain structure in Long Island Sound, where there is apparently a strong correlation between the observed stratification and contours of $h/u^3$ in September when the heat flux is small.

River discharge will be significant in modifying the vertical structure in this way when the freshwater buoyancy input is comparable to that due to surface heating in the seasonal cycle, i.e., for

$$(F/A)\Delta\rho \sim \alpha Q/c_p$$

where $\alpha$ is the expansion coefficient and $c_p$ is the specific heat of seawater. For a typical mid-latitude summer value of $Q \sim 100$ W/m$^2$ this implies

$F/A \sim 2$ cm/d. River runoff may exert particularly marked effects on the stratification in situations where the level of tidal mixing marginally exceeds the requirement for mixing down the seasonal heat input. Such a situation is that of Liverpool Bay, in the eastern Irish Sea, reported by Czitrom-Baus [1982].

Outflows from the rivers on the Lancashire coast, notably the Mersey, Dee, and Ribble, pro-duce stratification in a region of the bay where tidal mixing levels are just too great to allow significant thermal stratification ($\log_{10} \chi$ <2.7). Following periods of high runoff, haline stratification may be observed over a large area even during the winter months (Figure 11). Brack-ish water, produced by strong mixing near the estuaries, flows over the ambient water to produce a two-layer structure. In summer, this initial stratification suppresses vertical mixing and allows the accumulation of buoyancy input by heat-

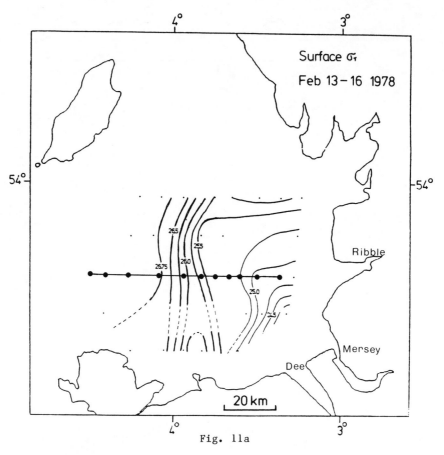

Fig. 11. Density structure in Liverpool Bay in the northern Irish Sea. (a)
Surface $\sigma_t$ distribution showing large E-W gradients close to 4°W on February 13-15,
1978. (b) East-west section along 53°49' showing strong salinity stratification to the
east of 4°W.

ing.  However, such stratification does not usually persist long because the water is relatively
shallow in this part of the Irish Sea and stratification may be rapidly eroded by the occurrence
of wind mixing.  The result is a highly variable
regime with short episodes of stratification separated by periods of almost complete vertical mixing, which is in marked contrast to the regular
seasonal cycle observed over most of the shelf.
At times of pronounced stratification, a sharp
front is frequently observed along a line close to
4°W where the two-layer structure breaks down.

The flow in the stratified area has many of the
characteristics of an estuarine circulation with
generally offshore flow in the surface layers.
The influence of rotation can, however, be seen in
the deflection of the current vector to the north
as in the case of coastal current discussed
above.  An analysis of the density-driven flow
under the influence of rotation in this area has
been given by Heaps [1972].

## 4.2.  Fronts Within Estuaries

It is only in the last few years that significant effort has been allocated to the study of the
relatively small scale frontal phenomena in estuaries.  The existing observations, which are
still too sparse to allow an adequate summary of
the overall importance of frontal effects in estuaries, have nonetheless served to identify several mechanisms operating to produce convergent
surface flows within estuaries.

4.2.1.  The tidal intrusion front.  The plume
fronts of the previous section may, in certain
circumstances, be forced back into the estuary by
the tidal inflow.  A clear example of this type of
tidal intrusion front has been reported recently
[Simpson and Nunes, 1981] for the small estuary of
the river Seiont in North Wales.  The freshwater
runoff in this short estuary is observed to flow
seaward as a density current whose velocity is
$\sim(g'D)^{1/2}$, where $g'$ is the reduced gravity and D

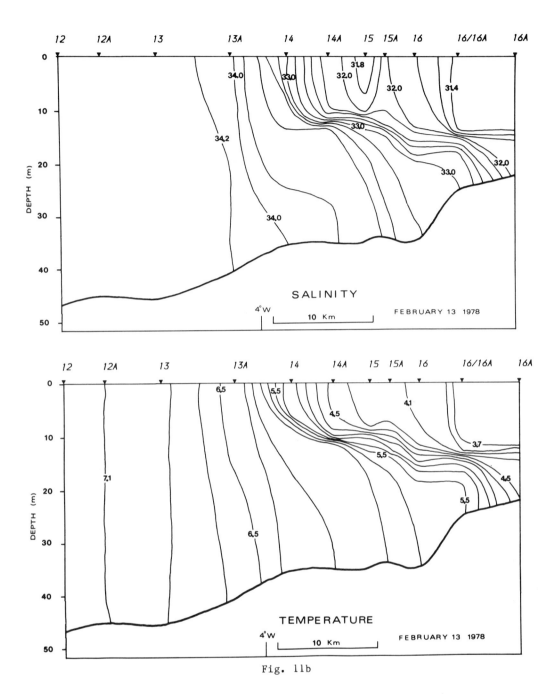

Fig. 11b

is the layer thickness (see Figure 12a). During the flood phase of the tide, the tidal inflow current may exceed this velocity, and the head of the gravity current retreats into the estuary. The strong convergence which occurs at the front or "plunge line" is two sided because mixing in the frontal zone results in the entrainment of surface layer fluid into an intermediate layer and this loss must be replenished by flow toward the front.

This and other properties of density current flows have been extensively studied in a series of laboratory experiments by Simpson and Britter [1979] and a simplified theoretical model has been formulated by Stigebrandt [1980]. A particularly interesting feature of the surface convergence observed in the Seiont is the V configuration (Figure 13) which it tends to adopt while within the estuary. At the apex of the V there is strong point convergence with an associated gyre system in which large quantities of surface material accumulate.

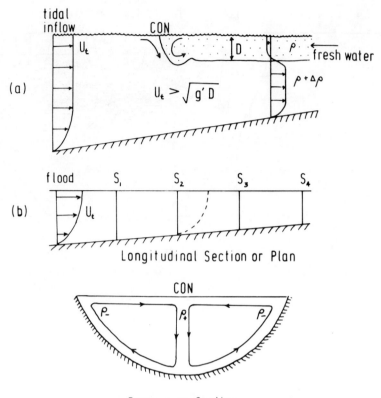

tidal
inflow       CON

(a)

$U_t$

$U_t > \sqrt{g' D}$

$\rho$ ← fresh water

$\rho + \Delta \rho$

D

flood    $S_1$    $S_2$    $S_3$    $S_4$

(b)

$U_t$

Longitudinal Section or Plan

CON

$\rho_-$     $\rho_+$     $\rho_-$

Transverse Section

Fig. 12. Mechanisms for frontogenesis in estuaries during the flood phase of the tide: (a) Freshwater layer propagating downstream as a density current which is arrested by the tidal inflow and forced back into the estuary around the time of maximum flood current. (b) Vertical and horizontal shear in the flood currents distorting the isohalines ($S_1$, etc.), in strongly mixed estuaries, to produce lateral gradients which drive a transverse circulation with strong convergence and sinking along the main current axis.

Evidence of similar tidal intrusion fronts has been obtained from other estuaries in North Wales, but in many cases the time scale of the fronts seems to be restricted to a short period on the flood before the increasing strength of the tidal stream causes a rapid erosion of the two-layer structure.

Examples of similar phenomena have been reported in fjord-type estuaries by Nunes [1982] for Loch Creran in Scotland and by Godfrey and Parslow [1976] for the Port Hacking estuary in New South Wales, Australia.

4.2.2. Convergence produced by transverse circulation. Fronts parallel to the axis of the channel occur in both stratified and mixed estuaries. In the stratified case, Bowman and Iverson [1978] suggest that the increased intensity of tidal mixing in shallow water breaks down stratification and, together with lateral shear in tidal flow, maintains a strong line convergence. This type of front, an example of which has been observed in Delaware Bay by Klemas and Polis [1977], is usually more pronounced under ebb than flood conditions.

Surface convergence is apparently maintained by mixing at the boundary, which produces an intermediate density water type with resultant sinking along the frontal interface.

A somewhat different type of front has been found to occur in some vertically mixed estuaries. Nunes [1982] reports a continuous line convergence extending ~10 km up the river Conway in North Wales (Figure 14). This feature occurs on the flood phase of the tide when conditions are vertically well mixed throughout most of the estuary.

The convergence in this case is associated with a transverse circulation which occupies the whole section (Figure 12b). Transverse density gradients are set up by the action of the horizontal and vertical shear in the tidal current on the longitudinal gradients. The pressure field associated with the transverse density structure then drives a two-celled circulation with a surface convergence along the axis of the channel. A diagnostic model of this process, based on observed density distributions, has been given by Nunes [1982]. It should be noted that horizontal grad-

Fig. 13. Tidal intrusion front in the estuary of the River Seiont, Caernarfon, North Wales. Vertically uniform saline water (~32%) is moving into the estuary from the right and sinking below a thin freshwater layer to form a two-layer structure on the left.

ients alone are sufficient to drive the circulation. Gravitational instability produced by the vertical shear has not been observed in the Conway, although it may occur in other estuaries as suggested by the measurements in Coos estuary on the Oregon coast by Burt and Queen [1957].

Velocities observed in the convergent surface flow are observed to be ~10 cm/s, or ~20% of the axial flow, which is much greater than the transverse flow due to secondary circulations of a homogeneous fluid in open channel flow [Gibson, 1909]. Further evidence for the proposed mechanism comes from the fact that this type of convergence is observed only on the flood and only when longitudinal salinity gradients are present. At times of extremely high runoff when all salt is flushed out of the upper estuary the convergence is not observed there.

### 5. The Modeling of Fronts

It is clear from these descriptions of fronts that they pose a formidable challenge to the analytic or numerical modeler. There is the problem of resolving sharp gradients, or even discontinuities, within a larger-scale circulation, and there is also the problem of incorporating incompletely understood, but important, physical processes. ın the following sections we consider fronts from the modeler's point of view and show how particular aspects of the topic have been tackled.

Models of fronts may be classified in several ways, for example by the method they use to take into account the sharp changes, whether by assuming a discontinuity (as in two-layer models) or by being able to resolve a large (but finite) gradient. Since fronts themselves may be classified into open ocean, boundary current, shelf break, upwelling, shallow tidally stirred sea or estuarine and coastal types, the models which are designed to simulate the different dynamics involved in these different cases may be similarly classified. Also, models have been devised to investigate particular physical processes in detail, such as the effect of variations in tidal mixing, double diffusion, cabbeling, frontogenesis, and instability.

Fig. 14. Axial convergence lines in the river Conway, North Wales, observed during the late stages of flood tide, July 6, 1982.

Here the layer models, which specify a pre-existing discontinuity, will be considered first. Then continuous models will be described. In each case there will be variations depending on the frontal type and the physics which the modeler has chosen to include.

Most of the models considered here are based on the following equations: these are the hydrostatic, Boussinesq, f-plane equations (see, for example, Phillips [1966]).

$$Du/Dt = fv - \partial\Phi/\partial x + F_u \qquad (2)$$

$$Dv/Dt = -fu - \partial\Phi/\partial y + F_v \qquad (3)$$

$$b = \partial\Phi/\partial z \qquad (4)$$

$$Db/Dt = F_b \qquad (5)$$

$$\partial u/\partial x + \partial v/\partial y + \partial w/\partial z = 0 \qquad (6)$$

where x, y, z are right-handed coordinates, z being measured upward from the reference sea surface level; (u, v, w) is the velocity; f is the Coriolis parameter, $\Phi = p/\rho_0$, where p is total pressure $P + \rho_0 gz$ and $\rho_0$ is the (constant) reference density; buoyancy $b = g(\rho_0-\rho)/\rho_0$, where $\rho$ is density, and $F_u$, $F_v$ and $F_b$ are frictional and mixing terms.

If temperature T and salinity S are to be considered separately, (5) is replaced by

$$DT/Dt = F_T + Q_T \qquad (7)$$

$$DS/Dt = F_S \qquad (8)$$

$$b = b(S,T,P) \qquad (9)$$

where $Q_T$ represents the effect of heating within the sea (for example, from penetrating solar radiation).

The most used but not necessarily correct (see, for example, Kirwan [1969]) choice for friction and mixing is to take eddy viscosities A and diffusivities K such that

$$F_u = \partial(A_H \, \partial u/\partial x)/\partial x + \partial(A_H \, \partial u/\partial y)/\partial y$$
$$+ \partial(A_V \, \partial u/\partial z)/\partial z \qquad (10)$$

$$F_v = \partial(A_H \, \partial v/\partial x)/\partial x + \partial(A_H \, \partial v/\partial y)/\partial y$$
$$+ \partial(A_V \, \partial v/\partial z)/\partial z \qquad (11)$$

$$F_b = \partial(K_H \, \partial b/\partial x)/\partial x + \partial(K_H \, \partial b/\partial y)/\partial y$$
$$+ \partial(K_V \, \partial b/\partial z)/\partial z \qquad (12)$$

with $F_T$ and $F_S$ similar. $A_H$, $A_V$, $K_H$, and $K_V$ may vary both in position and in time.

## 6. Two-Layer Models

### 6.1. Margules' Equation

The basic two-layer frontal model was presented over 100 years ago by Witte [1878] and also by Margules [1906]. Assume (see Figure 15) steady flow, no y variation, no friction, and no mixing. Then $fv = \partial\Phi/\partial x$ and $b = \partial\Phi/\partial z$. Hence v is independent of z in each layer, and the velocity difference across the interface is

$$v_1 - v_2 = (\rho_0 f)^{-1} g(\rho_2 - \rho_1)\partial h_1/\partial x \qquad (13)$$

This is known as Margules' equation. Note that v may still vary with x, and so may the interface and surface slopes. Hydrographic measurements in the vicinity of this front would give only the interface slope and hence only velocity differences, not the absolute velocity in each layer. As in all geostrophic flow calculations, the flow is indeterminate in the sense that there are an infinite number of solutions for v and $h_1$ which satisfy the equations and necessary boundary conditions.

### 6.2. Potential Vorticity

The flow becomes determined if we add other conditions. For example, consider potential vorticity. From the frictionless versions of equations (2) and (3) and from equation (6) it can be deduced that in a homogeneous layer of depth $h_1$,

$$D[(f + \omega)/h_1]/Dt = 0 \qquad (14)$$

where $\omega$ is the vertical component of vorticity, namely, $\partial v/\partial x - \partial u/\partial y$; $(f + \omega)/h_1$ is the potential vorticity; and equation (14) expresses the fact that this is conserved. Stommel [1958] provided a simple model of the Gulf Stream based on the assumption that potential vorticity is uniform, a condition that seems to hold fairly well according to observations. We then have $(f+\partial v/\partial x)/h_1 = f/h_0$, a constant, in the upper layer ($h_0$ is the depth scale of the baroclinic layer), and if the lower layer is assumed to be at rest, we have, from (13),

$$v_1 = (g'/f)\partial h_1/\partial x, \text{ where } g' = g(\rho_2 - \rho_1)/\rho_0$$

Hence $f + (g'/f)\partial^2 h_1/\partial x^2 = fh_1/h_0$, so

$$h_1 = h_0[1 - \exp(-x/R_d)] \qquad (15)$$

$$v_1 = (g'h_0)^{1/2} \exp(-x/R_d) \qquad (16)$$

where

$$R_d = (g'h_0)^{1/2}/f \qquad (17)$$

$R_d$ is called the "baroclinic Rossby radius of deformation."

A similar model was used by Csanady [1971] to determine the equilibrium shape of the thermocline in a shore zone, given the initial condition of a vertical interface between heated water near the shore and cooler water offshore: this determined the constant value of potential vorticity in each

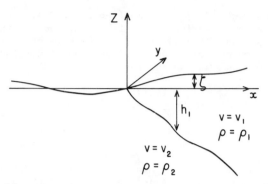

Fig. 15. The configuration of a two-layer frontal model.

layer (less in the heated region because of expansion of the water columns). He showed that either a wedge-shaped or a lens-shaped upper layer could be produced and also that with the initial condition of a horizontal thermocline interface an uptilt or downtilt of the thermocline was possible, with an exponential shape as in Stommel's [1958] Gulf Stream model. This theory was extended to take into account the effect of wind stress impulses by Csanady [1978].

### 6.3. Stability

Since meanders and eddies are ubiquitous in frontal regions, one should suspect that steady flow solutions are unstable, through either baroclinic instability (energy of the unsteady flow derived from available potential energy) or barotropic instability (energy of the unsteady flow derived from the kinetic energy of the mean flow). The general theory of such instability in quasi-geostrophic flows has been examined by Killworth [1980]. The stability of a Margules front confined between rigid horizontal boundaries has been studied by Orlanski [1968], and this model was extended by Flagg and Beardsley [1978] to the case of a sloping bottom. The results are given in terms of a Rossby number $Ro = \ell(v_1 - v_2)/2f$, where $\ell$ is the along-front wave number of the instability, and a Richardson number $Ri = g'h/(v_1 - v_2)^2$, where $h$ is the total depth. Orlanski [1968] showed that unstable waves exist at all wavelengths in the ranges of Ro and Ri that he examined, that is $0 < Ri < 5$, $0 < Ro < 3$. Flagg and Beardsley [1978] considered small Ro and so were able to use the semigeostrophic approximation [see Hoskins, 1975]. They concluded that while unstable modes do exist for finite bottom slopes, the e-folding times can increase substantially over those for the flat-bottom case, typically from 2-5 days to 50 days in the case of the shelf edge front south of New England and realistic shelf break slopes. This relative stability may explain why a persistent front is found near the shelf edge.

In all cases, one expects that the most unstable wavelength is of the order of a few times the Rossby radius of deformation (it is $3.9\ R_d$ in the classical Eady problem [Eady, 1949]). This can be shown by considering displacements of small parcels of fluid across the front. A parcel displaced at the same angle as the slope $\partial h_1/\partial x$ feels no restoring force, since it remains surrounded by fluid of the same density, while a parcel displaced horizontally feels a buoyancy force vertically, that is, at right angles to the direction of displacement. Displacement at an intermediate angle, however, will be amplified. In fact, the most unstable displacement direction is half the slope of the isopycnals in the case of continuous stratification. It can now be seen in a simple way how a bottom slope may stabilize the system, by reducing the range of directions in which such displacements may take place (at the bottom). On the other hand, if the bottom slope happens to be half the isopycnal slope, it may destabilize the system.

An analysis in terms of the potential vorticity equation (14), assuming small displacement velocities u and w, leads to a scaling

$$w/u \sim v\ell^2 h_0/f \sim 2Ro h\ell$$

For the most unstable wave,

$$w/u \sim \tfrac{1}{2}\,\partial h_1/\partial x = fv/2g' = f^2 Ro/\ell g'$$

Hence for the most unstable wave we have

$$2Ro h_0 \ell \sim f^2 Ro/\ell g'$$

so

$$\ell^2 \sim f^2/2g'h_0 = 1/2R_d^{\ 2}$$

This implies a wavelength $2\pi/\ell$ of the order of a few times the Rossby radius of deformation. Arguments of this type have been given by Orlanski and Cox [1973] and Pedlosky [1979]. It is interesting to note that this implies $Ro \sim Ri^{-1/2}$ [Pingree, 1979]. The time scale $R_d/v$ corresponds to approximately one third of the fastest e-folding time scale in the Eady problem and also appears to give a realistic growth time scale for frontal waves [Pingree, 1979].

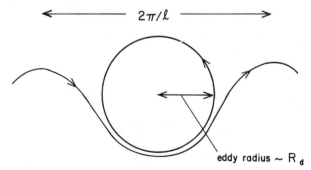

Fig. 16. Eddy size and the most unstable wavelength of a frontal boundary.

If the unstable waves eventually form eddies, the eddy radius will be of the order of $R_d$, as found by Gascard [1978] in the Mediterranean. This is shown in Figure 16. The problem of stability will be revisited in the context of continuous models in a later section. Typical values for the length and time scales in the case of shallow-sea fronts in mid-latitudes ($f \sim 10^{-4}$ s$^{-1}$) are as follows: $R_d$ = 5 km, velocity difference v = 0.1 m/s, doubling time scale a few days, Ro $\sim$ 0.1, Ri $\sim$ 25.

## 6.4. Friction

So far in the discussion of discontinuous models the assumption has been made that the main balance in equations (2) and (3) is between Coriolis and pressure gradient forces. This will be true if the Rossby number Ro (expressing the ratio of inertial to Coriolis forces, so of the form U/Lf, where U and L are characteristic velocity and length scales) and Ekman number E (expressing the ratio of frictional to Coriolis forces, so of the form $\upsilon/L^2 f$, where $\upsilon$ is the (eddy) viscosity) are both small. For the length scales appropriate to open ocean fronts, Ro and E are both in general small over most of the flow. However, near boundaries the length scale for velocity changes is reduced, and we have Ekman layers of thickness scale $(2\upsilon/f)^{1/2}$ in which E is not small. These allow the total solution to satisfy stress (or no-slip) boundary conditions which the interior geostrophic flow is unable to satisfy. Note that if depth L = 50 m, $\upsilon = A_V$ = 0.1 m$^2$/s (not unreasonable for shallow seas stirred by the tides), and f = $10^{-4}$, E $\sim$ 0.4, which is not small. The Ekman layer depth is in this case comparable with the total depth, and friction may not be neglected.

In the case where Ekman layers are much thinner than the depth, and if there is a significant stress across the interface produced by the geostrophic velocity difference, it can be seen (Figure 17) that this stress will drive an Ekman transport up the interface in the top (less dense) layer and down the interface in the bottom (denser) layer [Horne et al., 1978; G. T. Csanady, unpublished manuscript, 1972]. This in turn can produce a surface convergence in the denser layer and a surface divergence in the less dense layer. Garrett and Loder [1981] have derived

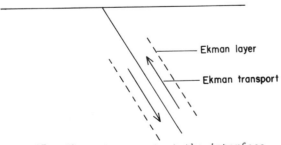

Fig. 17. Ekman transport at the interface.

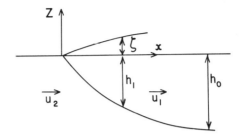

Fig. 18. A two-layer river plume front.

results for the lateral spread of the upper layer. If the interfacial stress is $\tau$, the mass fluxes in the interfacial Ekman layers will be $\tau/f$. Convergences and divergences in this Ekman flux produce changes in the depth of the upper layer, so (assuming the interfacial Ekman flux provides the only significant mass flux in the x direction) $\partial h_1/\partial t = \partial (\tau/f \rho_0)/\partial x$ while the upper layer is spreading. The stress $\tau$ depends on the velocity v'of the upper layer relative to the lower layer, which is $(g'/f)\partial h_1/\partial x$ according to Margules' equation (13). Hence a diffusion equation for $h_1$ is obtained. It is possible to consider the spread of a delta function, area

$$Q' = \int_{-\infty}^{\infty} h_1(x, \tau)\, dx$$

With an assumption of quadratic friction ($\tau = C\rho v'|v'|$) the front meets the surface at a finite value of x and with typical values of the parameters (C = 4 x $10^{-4}$, g' = $10^{-2}$ m/s$^2$, f = $10^{-4}$ s$^{-1}$, $h_1$ = 50 m at 10 km from the surface position of the front), Garrett and Loder [1981] deduce a rate of advance of the frontal interface at the surface of 8 cm/s.

This example demonstrates that one effect of friction is to introduce flow in the cross-frontal direction. In a two-layer model, there cannot be flow across the interface: this restriction is removed for continuous models (see section 7).

In the case of estuarine and river plume fronts the length scale L may be reduced to such a small value that the Rossby number and Ekman number are both large. Then Coriolis forces may be neglected altogether. Without rotation a steady state is possible only if the front advances at speed c relative to the denser fluid. Interfacial friction and mass entrainment then balance the pressure gradient in the x direction. Take coordinates moving with the front at speed c (Figure 18). This speed is of the order of the internal wave phase speed $(g'h_0)^{1/2}$. Consider a simple case in which the nonlinear terms may be neglected and the front propagates into still, deep water. We then have $u_2 = c$, and the equations of motion and continuity for the upper layer (from (2) and (6)) are

$$\partial \Phi_1/\partial x = A_V\, \partial^2 u_1/\partial z^2$$

(where $A_V$ is constant), with $\rho_0 \Phi_1 = \rho_1 g\zeta$ and

$$\int_{-h_1}^{0} u_1 \, dz = 0$$

This allows a solution $u_1 = A(h_1^2 - 3z^2)/3$, where

$$A = -(\rho_1 g/2A_V \rho_0)\partial\zeta/\partial x$$

if there is zero surface stress, where A is a function of x. A is determined by boundary conditions on the interface: the simplest is a no-slip condition such that $u_1 = c$ on $z = -h_1$. Then

$$A = -3c/2h_1^2$$

If the acceleration of the lower layer is negligible (as has been assumed by putting $u_2 = c = $ const),

$$\partial\Phi_2/\partial x = -g' \, \partial h_1/\partial x + (\rho_1/\rho_0)g \, \partial\zeta/\partial x$$

hence

$$\partial h_1/\partial x = 3A_V c/g' h_1^2 \qquad h_1^3 = 9A_V cx/g'$$

This gives the interface depth as a function of x. In the frame of reference moving with the front, velocities are in the sense shown in Figure 19. Officer [1976] and Bowman and Iverson [1978] give more details of this type of model. Garvine [1974] has presented a model with more realistic friction and entrainment boundary conditions. He also includes advective terms and uses depth-integrated equations. However, the model still relies on a number of assumptions, for example, a deep lower layer, steady two-dimensional flow, and specified velocity and density profiles, and contains several free parameters. This model was extended by Garvine [1979a, b] to include rotation. There are then two length scales, the dissipative length scale $h_0/C$ (where C is the interfacial friction coefficient) and the Rossby radius of deformation $R_d$. The frontal scale is the smaller of these two scales if their orders of magnitude are different. A hydraulic model for the nonrotating case, which conserves buoyancy, has been given by Stigebrandt [1980]. Garvine [1980] has developed his model to include advection and

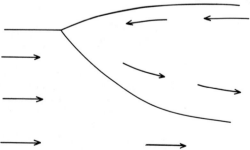

Fig. 19. Velocities in the frame of reference moving with the river plume front.

diffusion of buoyancy (equation (5)); this model indicates that mass entrainment is always downward, that fronts always spread relative to the ambient fluid, and that there is always a two-sided convergence in the cross-frontal flow. Garvine [1981, 1982] examines the problem of a buoyant discharge using jump conditions across the plume front similar to those in bore analysis.

## 7. Continuous Models

The two-layer models discussed above can describe the dynamics of fronts once they are set up in the form of a sharp discontinuity. They are, however, unable to simulate the generation and sharpening of fronts and the details of physical processes at the interface, for which we need to consider continuous models.

### 7.1. Frontogenesis

There are several processes which can produce fronts, and these may be listed as follows: (1) large-scale convergent flow (produced, for example, by variations in wind stress) in an area of initially weak horizontal temperature and/or salinity gradients, (2) frontogenesis (as in atmospheric frontogenesis) arising from instability and eddy formation in a flow in an area of initially weaker horizontal gradients, (3) variations in mixing, particularly tidal mixing in the case of shallow seas, which can produce a contrast between stratified and vertically well-mixed conditions within a short distance (the sharpness of this type of front may be enhanced by processes 1 and 2), (4) inflow of a different water mass, for example, in the case of estuaries and coastal waters, a flow of fresh water from the land (a front may be produced between the relatively fresh coastal water and the saltier seawater), and (5) upwelling (a possible explanation of some shelf edge fronts) in which wind stress causes a thermocline to reach the surface. Upwelling produced by the effect of variations in vertical eddy viscosity on wind-driven flow [see Heaps, 1980] is also a possible cause of shelf edge fronts. These frontogenetic processes will be considered in turn.

7.1.1. Large-scale convergent flow. The effects of a horizontal deformation field of flow ($u = -\gamma x$, $v = \gamma y$) on a basic density field independent of y were studied by MacVean and Woods [1980]. They assumed an inviscid fluid and so were able to use the semigeostrophic equations, which simplify by a transformation to geostrophic coordinates [Hoskins, 1975]: this assumes that the Rossby number (in the form $|D\underline{u}/Dt|/|f\underline{u}|$) is small, so that momentum is approximately geostrophic. The deformation rate $\gamma$ was chosen to be typical of that in mesoscale eddies in the deep ocean ($10^{-5}$ s$^{-1}$). A surface density discontinuity was predicted to be formed in 3-4 days. In the case where isotherms are inclined to isopycnals, it was shown that temperature fronts extend to greater depths than density fronts, since at these depths, flow tends to be along isopycnals.

7.1.2. Instability. The well-known property of the atmospheric polar front, that cyclones form on it and there is an intensification of temperature gradients (frontogenesis), producing "warm" and "cold" fronts, was modeled by Hoskins and West [1979], using the semigeostrophic equations referred to above. The values of the parameters used by Hoskins and West may be replaced by values appropriate for shallow seas: depth and velocity scales (h and U) are then reduced by at least two orders of magnitude. However the value of the Brunt-Väisälä frequency N (=(g'/h)$^{1/2}$) may be of the same order of magnitude. The upshot is therefore that the Rossby radius of deformation (and hence eddy radius) is measured in tens of kilometers in shelf seas rather than thousands of kilometers as in the atmosphere, but since time scales are nondimensionalized by dividing by Ri$^{1/2}$/f and since Ri (=N$^2$h$^2$/U$^2$) is of the same order of magnitude in the shelf seas and in the atmosphere, growth rate time scales are similar (a few days in mid-latitudes). There are analogues for the shelf seas of the diagrams given by Hoskins and West [1979]; of course, "surface" for the atmosphere becomes "bottom" for the sea, and "lid" becomes "surface."

The semigeostrophic equations do not include friction. As was remarked above, friction can be important in shallow seas subject to tidal stirring. A three-dimensional numerical solution of equations (2)-(6) was given by James [1981] for the case of a coastal front (in which the interface extends from surface to bottom) with an initially sinusoidal perturbation. This showed the development of a cyclonic eddy within 2 days, but eddy development could be suppressed by larger values of friction. The results showed entrainment of the denser water in the center of the eddy at the surface, frontogenesis, a movement of the eddy along the front, and evidence of the early stages of development of a cyclonic-anticyclonic vortex pair. Such vortex pairs are observed in the laboratory experiments of Griffiths and Linden [1981a, b, 1982] and in satellite images of fronts. James [1983] showed that after 6 days the model eddy had reached a mature form in which cross-frontal transfer had taken place and also that the eddy was associated with strong vertical motions driving fluid up and down the frontal boundary in the sense identified by Woods et al. [1977]. Similar results are obtained when this model is applied to a front with stratified water on one side, resembling the Norwegian coastal current. Figure 20 shows the relationship between the vortex pair (as seen in the steam function for the vertically integrated flow) and the surface buoyancy for a developing eddy in such a front. For more mature eddies the vortex pair lies on the stratified side of the front; Griffiths and Linden's [1982] experiments also indicate that eddy motions should propagate onto this side of the

front, when the upper layer is wide compared with the Rossby radius.

7.1.3. Variations in mixing. A sharp contrast in levels of stratification in shelf seas can be produced simply by variations in the effectiveness of tidal mixing from place to place. The model of James [1977] suggests that this effect alone can produce a contrast between strongly stratified and well-mixed conditions in a distance of 10-20 km (as shown in the density section of James [1978], derived from this model). The basic argument was put forward by Simpson and Hunter [1974] and states that a critical condition for vertically well-mixed conditions occurs when the rate of energy input required to mix the water completely over the depth h matches that supplied by turbulent mixing. This will be a fraction of the total rate of turbulent energy production by the stress.

Suppose there is a heat flux Q into the surface. The buoyancy flux is then $\eta = g\alpha Q/\rho c_p$, where $\alpha$ is the volume expansion coefficient of the water and $c_p$ is the specific heat of the water. If the reference level is at the bottom, potential energy per unit area is

$$V = \int_0^h \rho g z \, dz$$

If the water is well mixed, V= 1/2 $\rho$gh$^2$ = 1/2 Mgh, where M is the mass of water per unit area in the column. Hence

$$\partial V/\partial t = 1/2 \, Mg \, \partial h/\partial t = 1/2 \, \alpha g h Q/c_p = 1/2 \, \rho \eta h$$

This is the rate of energy input from the turbulence required to mix the water completely over the depth h. If the water stratifies, the potential energy will be less than that of the same column when well mixed. Now if the current is u, the rate of working of the bottom stress

$$\tau = \rho C u^2 = \rho u_*^2$$

where $u_*$ is the friction velocity, is $\tau u = \rho C^{-1/2} u_*^3$. If a fraction $\varepsilon$ of this ($\varepsilon$ is the "mixing efficiency") is available for mixing, the critical condition for complete mixing of the water column is

$$\varepsilon C^{-1/2} u_*^3 = 1/2 \, \eta h$$

or

$$C^{1/2} \eta h/2\varepsilon u_*^3 = 1$$

If C, $\eta$, and $\varepsilon$ are constant, this implies critical conditions at a specific value of h/u$^3$ (which in any case is clearly a key parameter).

Charts showing that many of the strongest fronts on the northwest European continental shelf lie along a contour of h/u$^3$ (where u is tidal current amplitude) have been presented by Simpson et al. [1977] and Pingree and Griffiths [1978]. The same has been done for the Bay of Fundy and Gulf of Maine by Garrett et al. [1978].

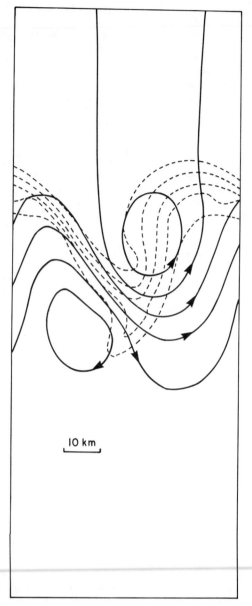

Fig. 20. Plan view of stream function for the depth-integrated flow (continuous lines: contour interval $2 \times 10^5$ m$^3$/s) superimposed on surface buoyancy contours in the frontal zone (dashed lines: contour interval $1.6 \times 10^{-3}$ m/s$^2$) in a model of a developing eddy in an idealized Norwegian coastal current. The less dense water is in the lower part of the figure.

The length scale $u_*^3/\eta$ is (when divided by von Karman's constant) the Monin-Obukhov length scale. It was pointed out by James [1980] that the critical value of $\eta h/u_*^3$ is of the order of 1, if the mean value of $u_*^3$ for neap tides is taken,

for the fronts considered by Simpson et al. [1977] and also that the critical value of the parameter for the onset of seasonal stratification in shelf seas is probably of the same order. Then the mixing efficiency is $\varepsilon \sim \frac{1}{2} C^{1/2}$, or 1-2% (this is somewhat higher than the estimates of Simpson et al. [1977] and Garrett et al. [1978] because the neap tidal current has been considered). At spring tides, stratification previously established at neap tides may reduce the mixing efficiency well below this value, so that the stratification is retained. This is a "feedback effect," which ensures that the tidal fronts do not move much between spring and neap tides (as predicted by James [1977] and observed by Simpson and Bowers [1979]), despite the large change in u$^3$.

The above argument ignores the details of the flow, since it is based on a depth-integrated energy balance. The effect of rotation, however, is to introduce Ekman layers of thickness scale $0.1 u_*/f$ [see Csanady, 1976], in which mixing effects tend to be confined (this was remarked on by Garrett et al. [1978]). As noted in a previous section, Ekman layers may be comparable with the depth in shallow seas, but when they are smaller than the depth, the parameter $fh/u_*$ may need to be considered in addition to $\eta h/u_*^3$. The mixing efficiency $\varepsilon$ may be expected to decrease with increasing $fh/u_*$.

The one-dimensional mixing model of James [1977] uses eddy diffusivities dependent on a local Richardson number, which are therefore reduced in stratified regions. This model reproduces the annual cycles of temperature for a number of positions in the vicinity of a front in the Celtic Sea, including the effect of the spring-neap tidal variation, and wind mixing.

All the one-dimensional vertical mixing models ignore the effect of horizontal advection and diffusion. Simpson [1981] has pointed out that advection by currents of only a few cm/s normal to the front should have a significant effect (see section 3.2). The success of the one-dimensional models in determining the position and structure of many tidal mixing fronts suggests that advection by wind-driven residuals (which may reach tens of centimeters per second) or tidal residuals (typically a few centimeters per second) has a minor influence, at least in the long-term mean. One explanation may be that cross-frontal advection is small because residual currents tend to flow along depth contours (except for wind-driven currents near the surface) and many fronts lie in the same direction.

7.1.4. _Inflow of water mass._ This is a case in which advection is important and was discussed above in the context of two-layer models of river plume fronts. A continuous model of a buoyant discharge has been given by Kao et al. [1977]. This was extended by Kao et al. [1978] and Kao [1980] to larger-scale oceanic fronts.

7.1.5. _Upwelling._ See the volume on coastal upwelling in this series [Richards, 1981].

## 7.2. Cross-Frontal Transfer and Details of Processes at the Interface

Cross-frontal circulation (which implies cross-frontal transfer) is predicted by diagnostic models, which assume a fixed density structure (near that observed) and derive the flow field from the equations of motion and continuity ((2), (3), (4), and (6)). The model of James [1978] assumes a stratified to well-mixed density transition similar to the summer front in the St. George's Channel area of the Celtic Sea and predicts upwelling on the well-mixed side of the front, together with a strong component of flow along the front. If the surface density change is large enough, there is a two-cell circulation in the vertical plane, with convergence near the front. This pattern was also found by Garrett and Loder [1981]. As Garrett and Loder point out, the strength of the predicted cross-frontal flow can imply that advection of the density field is important (that is, at least as large as the surface heating and mixing effects in the seasonal cycle model [James, 1977]). Therefore the assumed density structure would be slowly modified unless heating and mixing effects near the front are large enough to compensate for the advection (or unless the models overestimate the advective flow because assumed eddy viscosities are too large).

Garrett and Loder [1981] compare several mechanisms for cross-frontal transfer and suggest that this frictionally induced mean flow can be dominant. They show that for a small Rossby number, small Ekman number front with continuous stratification the depth of an isopycnal satisfies a horizontal diffusion equation, with (variable) horizontal diffusivity $(N^2/f^2)A_V$, and that the mean flow mass flux divided by the horizontal mixing flux resulting from a process parameterized by a horizontal diffusivity $K_H$ can be estimated by $(N^2/f^2)(A_V/K_H)$.

For the mechanism of shear dispersion (horizontal mixing resulting from a combination of vertical shear of horizontal currents and vertical mixing [see Fischer et al., 1979]) the appropriate value of $K_H$ is $(S^2/\sigma^2)K_V$, where $S^2$ is the mean square vertical shear of the horizontal current and $\sigma$ its frequency. Therefore the above ratio of fluxes is $(N^2\sigma^2/S^2f^2)(A_V/K_V)$, probably large in stratified conditions (the appropriate frequency $\sigma$ is that of tides or inertial currents).

Another mixing mechanism is that of barotropic eddies, generated by the interaction of tides and topography [Zimmerman, 1976, 1981], which may lead to high values of $K_H$ (up to $10^3$ m$^2$/s). This would dominate the mean flow flux but depends on the tidal and topographic conditions in the vicinity of the front. It is likely that eddies of this type may in certain regions influence the basic structure of a front, as in the case of headland fronts [Pingree et al., 1978].

Cross-frontal mixing by baroclinic eddies has already been mentioned in section 7.1: an esti-mate of this flux has been made by Pingree [1979] on the basis of Green's [1970] expression for poleward heat transfer in the atmosphere. This states that heat flux is proportional to $V_{max}\Delta\theta$, where $V_{max}$ is a maximum velocity resulting from release of potential energy and $\Delta\theta$ is the potential temperature difference. Pingree [1979] takes $V_{max} = (g'h_0)^{1/2}$ ($h_0$ is the depth of the baroclinic layer); hence the baroclinic eddy mass flux is $B(g'h_0)^{1/2}\Delta\rho h_0$. The constant B was taken by Green [1970] to be $5.5 \times 10^{-3}$ for the troposphere, but it is not clear immediately whether this is appropriate in shallow seas; it is related to the time scale of eddy formation, $T_e$. If an eddy radius R crosses the front every $T_e$ s, with one eddy passing in this direction every 4R m along the front, the mass flux is $\pi R^2\Delta\rho h_0/(4RT_e)$. Hence $B = \pi R/4R_d fT_e$. So if $R = R_d$, $B = \pi/4fT_e$: if $T_e$ is of the order of 10 days, $B \sim 10^{-2}$. If $K_H$ is the corresponding horizontal diffusivity,

$$h_0 K_H(\partial\rho/\partial x) = \pi R\Delta\rho h_0/4T_e$$

Now if $R = R_d$, $R = Nh_0/f$ (taking $N = (g'/h_0)^{1/2}$), so the ratio

$$(N^2/f^2)(A_V/K_H) = (N/f)(4T_e A_V/\pi h_0 L_s)$$

where $L_s$ is the horizontal front width scale. If $L_s \sim R_d$, this ratio is $\sim T_e A_V/h_0^2$. Now if $A_V = 10^{-2}$ m$^2$/s, $T_e = 10$ days, and $h_0 = 100$ m, this ratio is of the order of 1. That would imply that the cross-frontal mixing effectiveness of mean cross-frontal flow and baroclinic eddies can be comparable. A smaller value of $h_0$ would imply dominance of the mean flow flux. The ratio, of course, depends critically on the estimates of $T_e$ and $A_V$: if $T_e$ is increased, there is less exchange in eddies, and if $A_V$ is increased, there is a greater frictionally induced cross-frontal flow. The eddy horizontal mixing coefficient is $K_H \sim R_d^2/T_e \sim 10^2$ m$^2$/s if $R_d \sim 10$ km and $T_e \sim 10$ days. This is of the same order of magnitude as that estimated by Pingree [1979].

The final mechanism for cross-frontal exchange examined by Garrett and Loder [1981] was wind-driven transfer. If the wind stress is $\tau$, the wind-driven mass flux is of the order of $\tau\Delta\rho/\rho f$. This corresponds to $K_H = L_s\tau/\rho hf$. This is of the order of $10^2$ m$^2$/s (and so is comparable with the value for baroclinic eddies given above) if $L_s = 10$ km, $h = 100$ m, $f = 10^{-4}$ s$^{-1}$, and $\tau = 0.1$ Pa. This wind stress is produced by a moderate wind of about 8 m/s. Garrett and Loder [1981] note also that different values of $A_V$ on the two sides of a front may cause a convergence in the wind-driven flow at the front.

The conclusion can be drawn from this discussion (and the more detailed treatment of Garrett and Loder [1981]) that the mechanism of frictionally induced mean cross-frontal flow can be the most important cause of cross-frontal exchange. However, baroclinic eddies and wind stress could provide similar contributions. The particular circumstances determine which process, if any, dominates. Also, the Ekman number may not be small. A better estimate of the mean flow would probably be obtained from a coupled mean flow-frontal evolution model; a better understanding of viscosity and diffusivity parameters than exists at present would be desirable, since the mean flow depends critically on them (particularly $A_V$).

So far we have been considering mass transfer in fronts with a density contrast between the two sides. Many fronts, especially in the deep ocean, have little density change across them but have compensating temperature and salinity jumps. If isotherms and isohalines are inclined to isopycnals, density-compensating jumps in temperature and salinity tend to be produced, since motion tends to be along isopycnals, as in the MacVean and Woods [1980] model. In shallow seas, density changes are usually dominated by either temperature (as in well-mixed to stratified summer tidal mixing fronts) or salinity (as in near-coastal and estuarine fronts), and uniform density fronts have not been so generally reported.

Processes which become important in uniform density fronts are cabbeling, interleaving, and double diffusion. These have been studied by Garrett and Horne [1978], Horne et al. [1978], and Bowman and Okubo [1978].

Cabbeling describes the process of the mixing of two water masses of equal density but different temperature and salinity to produce a water mass denser than the original ones (this happens because of the nonlinearity of the equation of state (9)). Therefore downwelling is produced at the frontal boundary. Garrett and Horne [1978] show this downwelling velocity (in a vertically stratified zone) to be $w = -K_H \rho_{TT} (\partial T / \partial x)^2 / (\partial \rho / \partial z)$. For the example of a slope front off Nova Scotia, they deduce $w \sim 10^{-5}$ m/s (1 m/d). The downwelling produces a convergence at the surface, but this is insufficient in most circumstances to prevent the destruction of a front by horizontal diffusion. Bowman and Okubo [1978] model cabbeling in a front in unstratified water, and for their oceanic examples the maximum vertical velocities are $\sim 10^{-5} - 10^{-6}$ m/s, and the maximum horizontal velocities at the surface (which produce the convergence) are $\sim 10^{-3}$ m/s. However, their conclusions are sensitive to various unknown parameters.

Strong interleaving is often observed in uniform density fronts: this allows greater scope for mixing by increasing the surface area of contact of the two water masses [Joyce, 1977]. If the mixed water is swept away by some other process, the front can persist. Figure 21 shows the intrusion of a "hot salty" (HS) layer into "cold fresh" (CF) water. Double-diffusive effects [see Turner, 1973] then come into play; there is a diffusive interface above (where CF water overlies HS water) and a salt finger interface below (where HS water overlies CF water). At both interfaces there is an upward buoyancy flux, greater through the salt finger interface. Therefore an HS layer

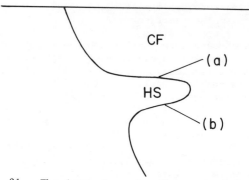

Fig. 21. The intrusion of a "hot salty" (HS) layer into "cold fresh" (CF) water, showing diffusive interface (a) and salt finger interface (b).

tends to rise and a CF layer to sink. The general upward buoyancy flux would produce a dip in the isopycnals at the front. The resulting pressure gradient then promotes a compensating upwelling at the front, at about 1 m/d for the case considered by Garrett and Horne [1978]. There would be a surface divergence. Garrett and Horne deduce, then, that the circulation driven by either cabbeling or double diffusion is small compared with that which would maintain or influence the overall structure of the front, although it is not much less than the estimated flow in the interfacial Ekman layers of a density front shown in Figure 17. The effectiveness of double-diffusive intrusions at fronts in promoting diapycnal (cross-isopycnal) mixing has been examined by Garrett [1982].

8. Discussion

Fronts are of profound importance in their influence on mixing processes in shallow seas. For example, the exchange of waters across a frontal boundary may be one of the most significant factors in determining how quickly a pollutant introduced at the coast is dispersed. Fronts are also usually the natural boundaries of the stratified areas of shelf seas. Therefore stratified models of circulation in shelf seas should include the effects of fronts; although, as is clear from this review, the subject of fronts has been a fruitful one for oceanographic modelers in recent years, there is still some way to go before this goal is achieved. Though much of the basic dynamics of fronts has now been explored (it lies in equations (2)-(9)), a factor limiting the rate of progress toward better models is ignorance of the correct parameterization of the terms $F_u$, $F_v$, and $F_b$. The frontal scale is one in a whole hierarchy of scales reaching from mean oceanic circulation to molecular dissipation ("10 decades of Fourier space" [Woods, 1980]). To model each scale requires parameterization of processes on scales too small to be resolved. In this sense the limiting problem (of turbulence parameterization) is common to the whole of physical oceanography.

Frontal models are designed to shed light on one part of the spectrum, a part which has a strong influence on the circulation, hydrographic conditions, and life in the seas around our shores. Improved models in the future should therefore ultimately lead to results of real predictive value concerning several important aspects of shelf sea oceanography.

## References

Argote, M. L. E., Perturbation of the density field by an island in a stratified sea, Ph.D. thesis, University of Wales, Bangor, 1982.

Aure, J., and R. Saetre, Wind effects on the Skaggerak outflow, in Proceedings of Symposium on the Norwegian Coastal Current, Geilo, September 1980, vol. 1, pp. 263-293, University of Bergen, Bergen, Norway, 1981.

Bowman, M. J., and W. E. Esaias, Fronts, stratification, and mixing in Long Island and Block Island sounds, J. Geophys. Res., 86, 4260-4264, 1981.

Bowman, M. J., and R. L. Iverson, Estuarine and plume fronts, in Oceanic Fronts in Coastal Processes, edited by M. J. Bowman, and W. E. Esaias, pp. 87-104, Springer-Verlag, New York, 1978.

Bowman, M. J. and A. Okubo, Cabbeling at thermohaline fronts, J. Geophys. Res., 83, 6173-6178, 1978.

Burt, W. V., and J. Queen, Tidal over-mixing in estuaries, Science, 126, 933-934, 1957.

Csanady, G. T., On the equilibrium shape of the thermocline in a shore zone, J. Phys. Oceanogr., 1, 263-270, 1971.

Csanady, G. T., Mean circulation in shallow seas, J. Geophys. Res., 81, 5389-5399, 1976.

Csanady, G. T., Wind effects on surface to bottom fronts, J. Geophys. Res., 83, 4633-4640, 1978.

Czitrom-Baus, S. P. R., Density stratification and an associated front in Liverpool Bay, Ph.D. thesis, University of Wales, Bangor, 1982.

Dessureault, J. G., "Batfish" - A depth controllable towed body for collecting oceanographic data, Ocean Eng., 3, 99-111, 1976.

Eady, E. T., Long waves and cyclone waves, Tellus, 1, 33-52, 1949.

Fischer, H. B., E. J. List, R. C. Y. Koh, J. Imberger, and N. H. Brooks, Mixing in Inland and Coastal Waters, 483 pp., Academic, Orlando, Fla., 1979.

Flagg, C. N., and R. C. Beardsley, On the stability of the shelf water/slope water front south of New England, J. Geophys. Res., 83, 4623-4632, 1978.

Floodgate, G. D., G. E. Fogg, D. A. Jones, K. Lochte, and C. M. Turley, Microbiological and zooplankton activity at a front in Liverpool Bay, Nature London, 290, 133-136, 1981.

Garrett, C. J. R., On the parameterization of diapycnal fluxes due to double-diffusive intrusions, J. Phys. Oceanogr., 12, 952-959, 1982.

Garrett, C. J. R., and E. P. W. Horne, Frontal circulation due to cabbeling and double diffusion, J. Geophys. Res., 83, 4651-4656, 1978.

Garrett, C. J. R., and J. W. Loder, Dynamical aspects of shallow sea fronts. Philos. Trans. R. Soc. London, Ser. A, 302, 563-581, 1981.

Garrett, C. J. R., J. R. Keeley, and D. A. Greenberg, Tidal mixing versus thermal stratification in the Bay of Fundy and Gulf of Maine, Atmos. Ocean, 16, 403-423, 1978.

Garvine, R. W., Dynamics of small-scale oceanic fronts, J. Phys. Oceanogr. 4, 557-569, 1974.

Garvine, R. W., An integral hydrodynamic model of upper ocean frontal dynamics, 1, Development and analysis, J. Phys. Oceanogr., 9, 1-18, 1979a.

Garvine, R. W., An integral hydrodynamic model of upper ocean frontal dynamics, 2, Physical characteristics and comparison with observations, J. Phys. Oceanogr., 9, 19-36, 1979b.

Garvine, R. W., The circulation dynamics and thermodynamics of upper ocean density fronts, J. Phys. Oceanogr., 10, 2058-2081, 1980.

Garvine, R. W., Frontal jump conditions for models of shallow, buoyant surface layer hydrodynamics, Tellus, 33, 301-312, 1981.

Garvine, R. W., A steady state model for buoyant surface plume hydrodynamics in coastal waters, Tellus, 34, 293-306, 1982.

Garvine, R. W., and J. D. Monk, Frontal structure of a river plume, J. Geophys. Res., 79, 2251-2259, 1974.

Gascard, J. C., Mediterranean deep water formation: Baroclinic instability and oceanic eddies, Oceanol. Acta, 1, 315-330, 1978.

Gibbs, R. J., Circulation in the Amazon River estuary and adjacent Atlantic Ocean, J. Mar. Res., 28, 113-123, 1970.

Gibson, A. H., On the depression of the filament of maximum velocity in a stream flowing through an open channel, Proc. R. Soc. London, Ser. A, 82, 149-158, 1909.

Godfrey, J. S., and J. Parslow, Description and preliminary theory of circulation in Port Hacking Estuary, CSIRO Div. Fish. Oceanogr. Rep., 67, 1-25, 1976.

Green, J. S. A., Transfer properties of large-scale eddies and the general circulation of the atmosphere, Q. J. R. Meteorol. Soc., 96, 157-185, 1970.

Griffiths, R. W., and P. F. Linden, The stability of vortices in a rotating stratified fluid. J. Fluid Mech., 105, 283-316, 1981a.

Griffiths, R. W., and P. F. Linden, The stability of buoyancy-driven coastal currents. Dyn. Atmos. Oceans, 5, 281-306, 1981b.

Griffiths, R. W., and P. F. Linden, Laboratory experiments on fronts, I, Density-driven boundary currents, Geophys. Astrophys. Fluid Dyn., 19, 159-188, 1982.

Heaps, N. S., Estimation of density currents in the Liverpool Bay area of the Irish Sea, Geophys. J. R. Astron. Soc, 30, 415-432, 1972.

Heaps, N. S., A mechanism for local upwelling along the European continental slope, Oceanol. Acta, 3, 449-454, 1980.

Holligan, P. M., Biological implications of fronts on the north-west European shelf, Philos. Trans. R. Soc. London, Ser. A., 302, 547-562, 1981.

Horne, E. P. W., M. J. Bowman, and A. Okubo, Cross-frontal mixing and cabbeling, in Oceanic Fronts in Coastal Processes, edited by M. J. Bowman, and W. E. Esaias, pp. 105-113, Springer-Verlag, New York, 1978.

Hoskins, B. J., The geostrophic momentum approximation and the semi-geostrophic equations, J. Atmos. Sci, 32, 233-242, 1975.

Hoskins, B. J., and N. V. West, Baroclinic waves and frontogenesis, II, Uniform potential vorticity jet flows - Cold and warm fronts, J. Atmos. Sci., 36, 1663-1680, 1979.

Ivanov, A., Oceanic absorption of solar energy, in Modelling and Prediction of the Upper Layers of the Ocean, edited by E. B. Kraus, pp. 47-71, Pergamon, New York, 1977.

James, I. D., A model of the annual cycle of temperature in a frontal region of the Celtic Sea, Estuarine Coastal Mar. Sci., 5, 339-353, 1977.

James, I. D., A note on the circulation induced by a shallow-sea front, Estuarine Coastal Mar. Sci., 7, 197-202, 1978.

James, I. D., Thermocline formation in the Celtic Sea, Estuarine Coastal Mar. Sci., 10, 597-607, 1980.

James, I. D., Fronts and shelf circulation models, Philos. Trans. R. Soc. London, Ser. A., 302, 597-604, 1981.

James, I. D., A three-dimensional model of shallow-sea fronts, in North Sea Dynamics, edited by J. Sundermann, and W. Lenz, pp. 173-184, Springer-Verlag, New York, 1983.

Joyce, T. M., A note on the lateral mixing of water masses, J. Phys. Oceanogr., 7, 626-629, 1977.

Kao, T. W., The dynamics of ocean fronts, 1, The Gulf Stream, J. Phys. Oceanogr., 10, 483-492, 1980.

Kao, T. W., C. Park, and H-P. Pao, Buoyant surface discharge and small-scale oceanic fronts: A numerical study, J. Geophys. Res., 82, 1747-1752, 1977.

Kao, T. W., H-P. Pao, and C. Park, Surface intrusions, fronts, and internal waves: A numerical study, J. Geophys. Res., 83, 4641-4650, 1978.

Killworth, P. D., Barotropic and baroclinic instability in rotating stratified fluids, Dyn. Atmos. Oceans, 4, 143-184, 1980.

Kirwan, A. D., Formulation of constitutive equations for large-scale turbulent mixing, J. Geophys. Res., 74, 6953-6959, 1969.

Klemas, V., and D. F. Polis, Remote sensing of estuarine fronts and their effects on pollutants, Photogramm. Eng. Remote Sensing, 43, 599-612, 1977.

MacVean, M. K., and J. D. Woods, Redistribution of scalars during upper ocean frontogenesis: A numerical model, Q. J. R. Meteorol. Soc., 106, 293-311, 1980.

Margules, M., Über Temperaturschichtung in sta-

tionär bewegter und ruhender Luft, Meteorol. Z., 1906, 241-244, 1906.

McClimans, T. A., Fronts in fjords, Geophys. Astrophys. Fluid Dyn., 11, 23-30, 1978.

Mork, M., Circulation phenomena and frontal dynamics of the Norwegian Coastal current. Philos. Trans. R. Soc. London, Ser. A, 302, 635-647, 1981.

Nunes, R., Dynamics of small-scale fronts in estuaries, Ph.D. thesis, University of Wales, Bangor, 1982.

Officer, C. B., Physical Oceanography of Estuaries (and Associated Coastal Waters), 465 pp., John Wiley, New York, 1976.

Orlanski, I., The instability of frontal waves, J. Atmos. Sci., 25, 178-200, 1968.

Orlanski, I., and M. D. Cox, Baroclinic instability in ocean currents, Geophys. Fluid Dyn., 4, 297-332, 1973.

Pedlosky, J., Geophysical Fluid Dynamics., 624 pp., Springer-Verlag, New York, 1979.

Phillips, O. M., The Dynamics of the Upper Ocean, 261 pp., Cambridge University Press, New York, 1966.

Pingree, R. D., Cyclonic eddies and cross-frontal mixing, J. Mar. Biol. Assoc. U.K., 58, 955-963, 1978.

Pingree, R. D., Baroclinic eddies bordering the Celtic Sea in late summer, J. Mar. Biol. Assoc. U.K., 59, 689-698, 1979.

Pingree, R. D., and D. K. Griffiths, Tidal fronts on the shelf seas around the British Isles, J. Geophys. Res., 83, 4615-4622, 1978.

Pingree, R. D., P. R. Pugh, P. M. Holligan, and G. R. Forster, Summer phytoplankton blooms and red tides along tidal fronts in the approaches to the English Channel, Nature London, 258, 672-677, 1975.

Pingree, R. D., M. J. Bowman, and W. E. Esaias, Headland fronts, in Oceanic Fronts in Coastal Processes, edited by M. J. Bowman and W. E. Esaias, pp. 78-86, Springer-Verlag, New York, 1978.

Proctor, R., Tides and residual circulation in the Irish Sea: A numerical modelling approach, Ph.D. thesis, University of Liverpool, Liverpool, England, 1981.

Richards, F. A. (Ed.), Coastal Upwelling, Coastal Estuarine Sci. Ser., vol. 1, AGU, Washington, D.C., 1981.

Richardson, K., H. F. Lavin-Peregrina, E. G. Mitchelson, and J. H. Simpson, Seasonal distribution of chlorophyll a in relation to physical structure in the western Irish Sea, Oceanol. Acta, 8(1), 77-86, 1985.

Savidge, G., A preliminary study of the distribution of chlorophyll a in the vicinity of fronts in the Celtic and western Irish seas, Estuarine Coastal Mar. Sci., 4, 617-625, 1976.

Simpson, J. E., and R. E. Britter, The dynamics of the head of a gravity current advancing over a horizontal surface, J. Fluid Mech., 94, 477-495, 1979.

Simpson, J. H., The shelf sea fronts: Implications of their existence and behaviour, Philos. Trans. R. Soc. London, Ser. A., 302, 531-546, 1981.

Simpson, J. H., and D. Bowers, Shelf sea fronts' adjustments revealed by satellite infra-red imagery, Nature London, 280, 648-651, 1979.

Simpson, J. H., and D. Bowers, Models of stratification and frontal movement in shelf seas, Deep Sea Res., 28, 727-738, 1981.

Simpson, J. H., and J. R. Hunter, Fronts in the Irish Sea, Nature London, 250, 404-406, 1974.

Simpson, J. H., and R. Nunes, The tidal intrusion front: An estuarine convergence zone, Estuarine Coastal Mar. Sci., 13, 257-266, 1981.

Simpson, J. H., D. G. Hughes, and N. C. G. Morris, The relation of seasonal stratification to tidal mixing on the continental shelf, Voyage of Discovery, Deep Sea Res. suppl. 327-340, 1977.

Simpson, J. H., C. M. Allen, and N. C. G. Morris, Fronts on the continental shelf, J. Geophys. Res., 83, 4607-4614, 1978.

Simpson, J. H., D. J. Edelsten, A. Edwards, N. C. G. Morris, and P. B. Tett, The Islay fronts: Physical structure and phytoplankton distribution, Estuarine Coastal Mar. Sci., 9, 713-726, 1979.

Simpson, J. H., P. B. Tett, M. L. Argote-Espinoza, A. Edwards, K. J. Jones, and G. Savidge, Mixing and phytoplankton growth around an island in a stratified sea, Cont. Shelf Res., 1, 15-32 1982.

Stigebrandt, A., A note on the dynamics of small-scale fronts, Geophys. Astrophys. Fluid Dyn., 16, 225-238, 1980.

Stommel, H., The Gulf Stream, 202 pp., University of California Press, Berkeley, 1958.

Turner, J. S., Buoyancy Effects in Fluids, 367 pp., Cambridge University Press, New York, 1973.

Wadhams, P., A. E. Gill, and P. F. Linden, Transects by submarine of the East Greenland Polar Front, Deep Sea Res., 26, 1311-1328, 1979.

Witte, E., Über Meeresströmungen, 45 pp., Pless (Schlesien) Verlag von A. Krummer, 1878.

Woods, J. D., Do waves limit turbulent diffusion in the ocean?, Nature London, 288, 219-224, 1980.

Woods, J. D., R. L. Wiley, and M. G. Briscoe, Vertical circulation at fronts in the upper ocean, Voyage of Discovery, Deep Sea Res., suppl., 253-275, 1977.

Wright, L. D., and J. M. Coleman, Effluent expansion and interfacial mixing in the presence of a salt wedge, Mississippi River Delta, J. Geophys. Res., 76, 8649-8661, 1971.

Zimmerman, J. T. F., Mixing and flushing of tidal embayments in the western Dutch Wadden Sea, II, Analysis of mixing processes, Neth. J. Sea Res., 10, 397-439, 1976.

Zimmerman, J. T. F., Dynamics, diffusion and geomorphological significance of tidal residual eddies, Nature London, 290, 549-555, 1981.

# THE COASTAL BOUNDARY LAYER AND INNER SHELF

N. R. Pettigrew

Ocean Process Analysis Laboratory, Department of Earth Sciences
University of New Hampshire, Durham, New Hampshire 03824

S. P. Murray

Coastal Studies Institute, Louisiana State University, Baton Rouge, Louisiana 70803

Abstract. Observational data from nearshore regions of the Great Lakes and from oceanic coastal zones of North America have shown that the coastal boundary layer (CBL) is an environment which differs markedly from offshore regions. A combination of factors including the presence of the coastal boundary, depth variations, and both horizontal and vertical density gradients are shown by theoretical and observational studies to exert strong influence upon the hydrographic and hydrodynamic character of the CBL. Evidence strongly suggests that nearshore currents are fundamentally three dimensional in character and are dependent upon upcoast, as well as local, forcing.

## Introduction

The coastal boundary layer is a unique fluid-dynamical regime found near the shores of large lakes, shallow seas, and continental shelves. This regime often displays kinematic, dynamic, and hydrographic characteristics which differ markedly from those observed farther from shore.

The first clear distinctions between nearshore and offshore conditions were recognized nearly two decades ago in investigations of the North American Great Lakes. Many investigators noted that large thermocline displacements associated with upwelling and downwelling cycles were primarily confined to a nearshore band of roughly 10 km extent. In addition, Verber [1966] noted that motions near the shoreline of Lake Michigan were characteristically along-isobath, while midlake currents were dominated by rotary oscillations of near-inertial frequency. Early theoretical insights into some of the important physical mechanisms responsible for these contrasting flow features were provided by the analytical baroclinic coastal jet models of Csanady [1967, 1968] and Birchfield [1969].

In a pair of landmark papers analyzing detailed coastal zone measurements in Lake Ontario, Csanady [1972a, b] first used the term coastal boundary layer (CBL) to describe the nearshore band over which midlake motions "adjust to the presence of the shores." He also discussed several of the major physical factors, and their mathematical counterparts, which lead to the boundary layer character of the shore zone.

These investigations of the Great Lakes formed the foundation of the coastal boundary layer as a conceptual model. Although in the oceanic case the nearshore response is influenced by tides and the larger-scale flow of the continental shelf, many of the mechanisms originally identified in lakes are found to have powerful influence on the dynamics of the oceanic coastal boundary layer as well. Of the many factors which influence motions nearshore and serve to differentiate the shore zone from offshore regions, some of the most important are listed below.

1. The coastal boundary blocks flow perpendicular to the coast at the shoreline. This is the fundamental constraint to which any flow must adjust within the CBL.

2. The seaward deepening of the water column results in a topographic boundary layer through the mechanism of vortex stretching in association with cross-isobath flow.

3. The shallowness of the nearshore water column enhances the influences of frictional forcing and dissipation.

4. The seasonal occurrence of strong stratification in shallow water leads to dramatic cross-shore variability (upwelling, downwelling, and coastal jets) over length scales of the local baroclinic deformation radius.

5. The influence of runoff and preferential vernal heating nearshore dramatically alters the hydrographic structure of the CBL and results in thermohaline flow features.

Each of the mechanisms listed above contributes to the boundary layer character of the shore zone, and each possesses its own characteristic decay scale. It is easily appreciated how the interplay of these and other effects may lead to a complex and variable structure involving several imbedded layers within the zone constituting the CBL/inner-shelf system. The distinction sometimes drawn between the coastal boundary layer and the inner shelf is somewhat artificial, and the nearshore zone is better thought of as a smooth transition between flow regimes dominated by some combination of the constraints listed above (with characteristic cross-shore scales of roughly 10 km) and the shelf regime subject to much larger-scale influences.

In the shallowest waters of the coastal region, within the order of 100 m of the beach, is the wave-driven surf zone. In this region the principal source of momentum flux is the so-called radiation stresses [Longuet-Higgins and Stewart, 1962] exerted by shoaling waves. The generation of alongshore currents in the surf zone by obliquely incident waves has been understood for some time [Bowen, 1969; Longuet-Higgins, 1970a, b]. This process is far more important than any direct driving by wind stress.

The relationship between the wave-driven regime and the classical CBL remains an unsettled issue. It has recently been suggested that wave-induced momentum flux may generate significant currents outside the surf zone [Curtis, 1979; Dolata and Rosenthal, 1984], which would be superimposed upon the wind-driven, thermohaline, and pressure gradient flows of the CBL/inner shelf.

The purpose of this article is to discuss briefly some of the simple theoretical ideas about the CBL which have evolved over the past several years and to try to gain some insight into the important observational data available. The data are discussed in a geographic context, since unifying physical processes have not yet been fully delineated.

## Theoretical Concepts

Analytical models of the CBL have tended to be highly idealized linear representations of some combination of the five principal factors listed above. Two-dimensional barotropic numerical models [Bennett, 1974; Bennett and Magnell, 1979] show that the neglect of nonlinear cross-shore advection is probably not a serious limitation, although they suggest that the alongshore advection could be significant. Perhaps the chief strength of CBL models has been their relative simplicity, which has allowed straightforward evaluation of the effects of the physical constraints and processes included. Their intent and utility are limited to providing general insight into the physics of the CBL rather than providing detailed quantitative predictions. Our treatment of these modeling efforts will necessarily be very limited in both scope and detail, and discussion will be confined to important concepts or models which are representative of the findings of many investigators.

## The Coastal Boundary and Stratification

Although all models of nearshore circulation must take account of the coastal constraint as a boundary condition, the presence of a coast does not, by itself, lead to differentiation between nearshore and shelf flow conditions. In a flat-bottom, homogeneous coastal zone with no bottom friction, theoretical solutions are scaled by the external Rossby radius of deformation, which is generally greater than the full shelf width even for shallow (20 m) coastal zones. On the other hand, a stratified system of the same coastal geometry has rotationally trapped solutions which have the much smaller baroclinic deformation radius as a characteristic cross-shore decay scale. In the Great Lakes and in the CBL south of Long Island, New York, this length scale is often calculated to be roughly 5 km [e.g., Csanady, 1977; Pettigrew, 1981], suggesting that the combination of rotation and stratification in even the most idealized shore zone can lead to motions effectively confined to within 5 to 10 km of the coast.

This mechanism has been investigated in detail by Csanady [1976, 1968, 1977] in the context of a two-layer fluid bounded by an infinite straight coast and a flat bottom without dissipation. The aperiodic part of the response of this system to an impulsive alongshore wind stress is the well-known coastal jet shown schematically in Figure 1. This solution is characterized by an alongshore velocity which is directly proportional to the wind stress impulse (wind stress times duration), has an offshore e-folding scale of the baroclinic deformation radius, and has its near-shore momentum confined primarily to the upper layer. Seaward of one deformation radius, the alongshore flow is approximately equal in both layers, while far from shore the bottom layer is motionless and the top layer is characterized by Ekman drift to the right of the wind. Solutions for the displacement of the density interface show upwelling and downwelling cycles also confined principally to the CBL. The details of these solutions as well as those for cross-shore winds are given by Csanady [1977], and the wavelike transient responses of the surface, interface, and currents have been discussed by Crepon [1967, 1969] and Pettigrew [1981]. All of these models make clear the dominant role of the alongshore wind stress in the generation of motions in the coastal boundary layer.

A chief limitation of the coastal jet model is that it does not include dissipation, so that in the case of alongshore wind stress forcing, the solution is only valid during an initial period (usually several inertial cycles, depending on the wind stress impulse) after which the linearization and neglect of bottom and interfacial friction are no longer justified. The coastal jet model was generalized by Allen [1973] to include continuous stratification and eddy diffusion of momentum.

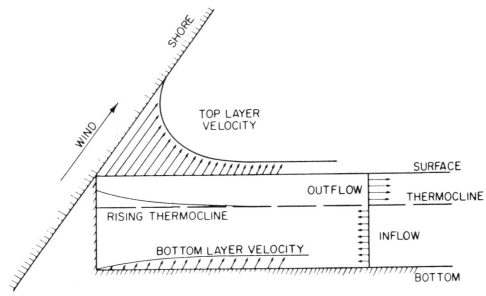

Fig. 1. Schematic picture of coastal jet development in a two-layer fluid without bottom topography [from Csanady, 1977].

Allen found that a coastal jet developed within a few days but that a final diffusive steady state took much longer to achieve. Despite its simplicity the coastal jet conceptual model has been remarkably successful in providing insight and reproducing coastal flows and upwelled features of the proper order of magnitude.

Bottom Topography

The effects of depth variatons upon coastal flows may be conveniently separated into direct effects due to kinematic constraints and differential acceleration of the water column by surface (as opposed to volume) forces, and indirect effects due to the combined influence of topography and the rotation of the earth. These mechanisms are all particularly important in the CBL, where both the bottom slope and the proportional depth variations are characteristically large.

The effect of a sloping beach upon the two-layer coastal jet was discussed by Csanady [1977], who gave an approximate solution for a basin with a flat bottom and linearly sloping sides. The solution, which is valid only where the total depth exceeds the top-layer equilibrium depth (i.e., the two-layer portion), shows interesting, although not marked, differences from the simple flat-bottom case. The offshore decay scale of the sloping shoreline model is given by $[(h_1 + h_2)/h_2]^{1/2}$ times the flat-bottom baroclinic deformation radius, where $h_1$ and $h_2$ are the depths of the upper and lower layers, respectively. This result suggests that the trapping width of the coastal jet over topography will be increased relative to

the corresponding flat-bottom case. When the bottom-layer depth greatly exceeds the top, the magnitude of this boundary layer broadening is negligible.

The principal modification to the structure of the coastal jet is a more even nearshore distribution of momentum in the two layers, although the upper layer is still favored in this regard. The source of increased alongshore momentum in the bottom layer is a rotational effect upon the cross-shore flow associated with an upwelling or downwelling thermocline over a sloping shore. Thermocline displacements over a sloping bottom require cross-shore flow in the bottom layer equal to the vertical velocity divided by the bottom slope. This cross-shore velocity in turn produces an alongshore acceleration via the Coriolis effect.

A more profound effect of topography is manifest in the barotropic flow component, which may acquire topographically controlled decay scales of the order of 10 km. This barotropic component, not present in flat-bottom systems, arises from wind momentum being distributed over a water column of variable depth. In terms of the depth-integrated vorticity balance, one may think of the curl of the wind stress divided by water depth as the relevant forcing function which must be balanced by vortex stretching, changes in the relative vorticity (horizontal shear), or friction.

The presence of depth variations allows the possibility of topographic or shelf waves. As shown by Longuet-Higgins [1968] and Brink [1983], vorticity waves may propogate along submarine topographic features and do not require a coast to exist. However, as pointed out by Gill and Schumann [1974], the coastal boundary constraint is of great importance in the wind generation of topographic waves in regions deeper than the depth of direct penetration of the wind-imparted momentum. In the CBL the existence of steep bottom slopes, a coastal boundary, and a shallow water column all suggest an environment conducive to the generation and propagation of vorticity waves. Clear and detailed evidence of the importance of such wavelike motions has been presented for the coastal boundary layer in the Great Lakes [e.g., Csanady and Scott, 1974; Clarke, 1977]. The fact that parallel evidence has not been found for the oceanic CBL [Csanady, 1982] could be due in part to experimental design. Field investigations concentrating on the first 10 km from shore have generally not been designed with alongshore arrays for sensing wave propagation. However, it is also possible that the existence of strong tidal currents and nearshore dissipation limits the importance of free waves in the marine CBL. Simons [1983] suggests that topographic waves do play an essential role in the barotropic response of nearshore currents to wind, but their effects are difficult to detect in the time domain momentum-budget approach often used in analysis of nearshore velocity data.

## Topography and Friction

Large proportional depth changes in the coastal zone can play an extremely important role in determining the nearshore response. In a shallow homogeneous system it is clear that the current speed arising from a particular wind stress impulse will be inversely proportional to the water depth over which the momentum is distributed. This effect suggests that wind stress becomes the dominant forcing as the water column shoals. As pointed out by Scott and Csanady [1976], as the beach is approached, wind stress overwhelms pressure gradient forces so that eventually a force balance between wind stress and dissipation may be expected. One effect of bottom topography is thus the elimination of static setup as a possible response to steady wind over an enclosed or semi-enclosed basin. As shown by Csanady [1973] for an oblong basin with depth variation and no dissipation, wind stress and pressure gradient forces balance one another only at locations where the local depth equals the average depth of the basin. Shoreward of such locations, wind stress predominates, and coastal currents are accelerated downwind, while in deeper regions a compensating return flow is driven by the pressure gradient force (proportional to the total depth). This situation leads to so-called "topographic gyres." Similar effects may be noted along open coastlines where local and externally imposed pressure fields may oppose the wind [e.g., Csanady, 1978].

In addition to increasing the influence of wind stress forcing, shallow depths of the CBL also make bottom stress and dissipation an important element in the nearshore momentum balances. In this regard, the large alongshore acceleration and transports predicted by linear inviscid models are unrealistic in very shallow regions. After an initial period of transient acceleration, one might expect a frictional adjustment as wind stress and bottom stress approach equilibrium. By invoking the coastal constraint and assuming a long, straight coast with arbitrary cross-shore depth variation and quadratic bottom stress, Csanady [1982] calculated a frictional adjustment time given by

$$T_f = \frac{H(x)}{2u_* C_d^{1/2}}$$

where $H(x)$ is the water depth, $C_d$ the drag coefficient referred to the depth-averaged velocity, and $u_*$ the so-called friction velocity equal to the square root of the wind stress divided by the density of seawater. For a typical wind stress of 1 $dyn/cm^2$ and a drag coefficient of $2 \times 10^{-3}$, one calculates an adjustment time just over 9 hours for a water depth of 30 m. These results suggest that in the coastal boundary layer, frictional adjustment to strong wind takes place within a few hours and the flow quickly represents a quasi-steady state. During a wind stress episode an inner region representing a nearshore frictional regime extends seaward into deeper water as time passes. In response to light winds the CBL may take several days to reach a frictional equilibrium velocity ($u_* C_d^{-1/2}$) representing a balance between wind stress and bottom stress.

As a first step toward understanding quasi-steady flow in the oceanic CBL, Scott and Csanady [1976] considered a simplified equilibrium model. The model assumes that the coastal constraint prohibits depth-integrated cross-shore flow everywhere within the CBL, which in turn limits alongshore variation to a constant pressure gradient. After neglecting cross-shore stresses relative to the Coriolis force associated with alongshore currents, the cross-shore momentum equation is geostrophic, and the alongshore equation is a balance between wind stress, linear bottom stress, and the alongshore pressure gradient. As pointed out by Bennett and Magnell [1979] and Noble and Butman [1979], the alongshore pressure gradient along the east coast of the United States is actually a strong function of both time and space. Smith [1979, 1980a] and Pettigrew [1981] have shown that cross-shore transport is often substantial within the CBL, and alongshore flow variations are therefore required in a consistent view of nearshore circulation. Nevertheless, despite the overly restrictive constraints, Scott and Csanady's [1976] model succeeded in making clear the important roles of friction and the alongshore pressure gradient and provided a means of estimating their orders of magnitude in the CBL environment.

## Topography, Friction, and Alongshore Variations

Many of the important effects of bottom topography, dissipation, and longshore variability have been included in an illuminating model often referred to as the arrested topographic wave [Csanady, 1978]. In this steady state model, water depth varies with cross-shore position, and bottom friction is a linear function of the depth-averaged alongshore velocity. In addition, non-zero depth-averaged cross-shore velocity is retained, and the pressure, velocity components, and wind stress are allowed to vary in the alongshore direction. The alongshore dependence of the wind stress is considered to arise not only from spatial variation in the wind field but also from changes in coastal orientation. In this way, major coastline changes are taken into account insofar as they affect the local alongshore and cross-shore components of the wind stress, while much of the analytical simplicity of straight-coast models is retained.

The vorticity dynamics of the model represent a balance between the curl of the ratio of net stress (wind stress minus bottom stress) to water depth and a vortex stretching term associated with cross-isobath flow. In this sense the system of equations may be thought of as a shallow-water topographic analogy to the well-known Sverdrup relation. The important point to recognize is that the arrested topographic wave represents a barotropic frictional boundary layer which would not exist in the absence of a sloping bottom.

Wind-driven solutions of the arrested topographic wave model are characterized by significant cross-shore transport along coastal sections acted upon by alongshore winds. This net cross-shore transport is balanced at other coastal locations either by the downcoast (the direction in which free topographic waves propagate) seaward broadening of the shore-trapped circulation or by cellular return flow in the case of spacewise-periodic variations in the wind field or coastline orientation. The integrated alongshore transport at a particular alongshore location is determined by the net cross-shore transport occurring upcoast of that location.

The arrested topographic wave model has proven to be a very useful aid in understanding the importance of alongshore variability in the wind-driven response of a sloping shore zone. It also provides insight into the process by which quasi-steady alongshore pressure gradients may be generated and maintained by spatial variation of wind and coastal geometry. The offshore trapping scale of the barotropic circulation over a linearly sloping bottom may be expressed as

$$L_x = \left(\frac{rL_y}{fs}\right)^{1/2}$$

[cf. Winant, 1979], where r is a bottom friction coefficient with dimensions of velocity, $L_y$ is the alongshore length scale determined by the wind field or by variation in coastal geometry, f is the Coriolis parameter, and s is the bottom slope. Clearly, the specific cross-shore scales of the arrested topographic wave response are strongly dependent upon a variety of geographical factors, and several such boundary layers may be superimposed. Typical values of $L_x$ are 10 to 30 km, and the circulation has its strongest influence on the CBL and inner shelf.

What is now recognized as an arrested topographic wave solution was earlier discussed by Pedlosky [1974] in the context of viscous boundary layer expansion for a closed basin. Among the complex structures of a stratified fluid with sidewall friction layers, Pedlosky found a barotropic "topographic boundary layer" dynamically identical to the circulation cells discussed above. This consistency is very reassuring in view of the simple depth-integrated equations used in Csanady's [1978] derivation.

The arrested topographic wave model was specifically adapted for the coastal boundary layer/inner shelf by Pettigrew [1981]. In this work it was noted that relatively modest differences in the analytical topography were capable of inducing significant changes in the detailed structure of the theoretical nearshore pressure and flow fields. Accordingly, solutions were obtained for a more general form of bottom topography which is capable of accurately modeling the depth distribution of the CBL/shelf system in a wide variety of geographical locations.

## Thermohaline Forcing

The CBL and inner shelf can be strongly influenced by freshwater input and local heating and cooling. These processes generate horizontal density gradients which in turn drive a thermohaline circulation and alter the nearshore response to other forcing.

Heating or heat extraction tends to be rapid and effective within the CBL because of the relatively small heat capacity of the shallow water column. This situation leads to horizontal density contrasts arising between the CBL and deeper regions, even under the influence of spatially uniform atmospheric inputs. The effect of simple thermal expansion due to heating of a vertically mixed nearshore water column would be an alongshore current in the downcoast direction which is maximum near the surface and zero at the bottom. Thus preferential nearshore warming (or cooling) can cause net alongshore transport.

The effects of freshwater runoff on nearshore circulation are similar to the effects of heating, although the magnitude of the water masses involved is liable to be greater along coastlines with major riverine inputs. Depending upon the volume of freshwater runoff and the coastal geometry, the entire shelf may be affected.

The development of a hydrographic boundary layer in response to freshwater discharge at the coast has been investigated in a two-dimensional numerical model by Kao [1981]. Solutions are characterized by realistic frontal systems and density-driven baroclinic jets which flow in the downcoast direction. Kao's work emphasizes the dynamical importance of the ageostrophic cross-shore transport. Outside the bottom Ekman layer in steady state, the Coriolis force associated with cross-shore flow is balanced by approximately equal viscous and nonlinear momentum fluxes.

In addition to cross-shore density gradients already discussed, estuarine or riverine "point sources" also cause quasi-steady alongshore density gradients. The role of alongshore variation in the density field in continental shelf regimes has been discussed by Beardsley and Hart [1978], Hsueh and Peng [1978], and Csanady [1979]. Some of these effects may be qualitatively understood in terms of the vorticity balance discussed earlier in connection with the arrested topographic wave model. The alongshore density gradient (in the vertically mixed model) implies a cross-shore geostrophic flow component with consequent vortex stretching, which must be balanced by the curl of the bottom stress divided by the water depth. In this way the action of the alongshore density gradient is analogous to the wind stress curl and produces circulation in addition to that driven by wind and cross-shore gradients.

Csanady [1984] has shown that along-isobath density gradients may generate depth-independent pressure field and circulation components through a density gradient/topographic interaction. It is suggested that this mechanism could make a significant contribution to the mean flow in low-latitude coastal regions subject to high runoff.

## CBL/Inner-Shelf Observations

### The Great Lakes

The observation of nearshore circulation in the Great Lakes is simplified by the absence of tidal currents which add a large "noise" component to oceanic coastal observations. The lack of tidal oscillations made particularly clear the contrast between the nearshore parabathic flow and the midlake rotary motions described by Verber [1966] in Lake Michigan. These data were analyzed in detail by Malone [1968], who reported that nearshore inertial energy levels were reduced by roughly a factor of 4 relative to midlake values. He also noted that near-inertial currents occurred only in the presence of a thermocline and that they exhibited 180° phase difference between upper and lower layers. Smith [1972] observed similar behavior in Lake Superior. These characteristics may be attributed to the influence of the coastal constraint [Pettigrew, 1981].

The nearshore zone was early recognized as a thermal boundary layer due to enhance shallow-water heating and mixing processes which often lead to nearshore and offshore zones being separated by a thermal front [Rodgers, 1965; Csanady,

1971]. As mentioned earlier, the existence of this type of density structure is indicative of boundary currents, a well-known example of which is the Keweenaw Current in Lake Superior [Niebauer et al., 1977].

The most comprehensive investigation of the Great Lakes was carried out in Lake Ontario in connection with the International Field Year on the Great Lakes (IFYGL). Because of the evidence that the first 10 km or so from the coast represented a regime distinct from the interior, detailed nearshore investigations were carried out at five locations around the lake.

The bulk of the observations collected in this field experiment consist of quasi-synoptic current and temperature transects across the CBL. Csanady [1972a, b] described the spatial and temporal structures of the coastal currents and associated upwelling and downwelling cycles for the spring and summer-fall seasons. During the spring season, mean currents and wind response were generally low. Significant flow was confined primarily to an approximately 5-km-wide band of warm water in the shallow nearshore region. The confinement of flow to this band was attributed to a combination of rotational trapping and the increased transfer of wind momentum over the warm water (reduced atmospheric stability). In summer-fall the stratification was much stronger and extended across the lake. The response to wind was greater, as was the total kinetic energy. Baroclinic wavelike motions were apparent offshore, while persistent alongshore currents and associated thermocline tilts were trapped within roughly 7 km of shore. Csanady [1972a, b] emphasized throughout these discussions that the CBL should be viewed as a physical environment rather than a boundary layer solution to a particular dynamical balance.

Moored current meter data from locations 3, 6, and 11 km from the shore at Oshawa, Lake Ontario, during IFYGL were analyzed by Blanton [1974]. He concluded that the total kinetic energy decreased with distance from shore during all seasons and that during summer stratified conditions a marked transition from predominantly rectilinear flow to predominantly rotary motion took place somewhere between the moorings located 6 and 11 km from shore. This transition is due to the influence of the coastal constraint, which, under stratified conditions, is approximately scaled by the baroclinic deformation radius.

More recent observations by Murthy and Dunbar [1981] in Lake Huron showed a more detailed structure within the CBL. Using data from a very closely spaced mooring array (nine moorings within the first 8.6 km from shore), Murthy and Dunbar found evidence of two distinct zones within the CBL, each of which exhibited classic boundary layer characteristics. As shown in Figure 2, mean velocity components were observed to peak between 2 and 3 km from shore. The region shoreward of this flow maximum is interpreted as an inner, frictional boundary layer within which the flow is

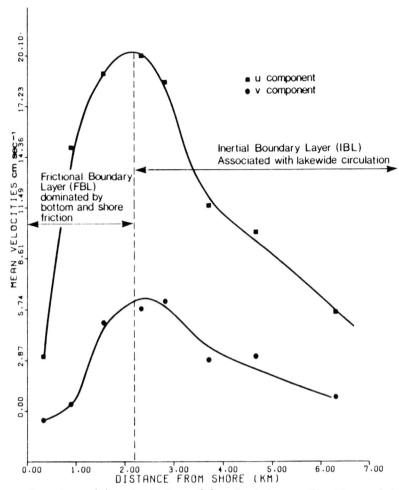

Fig. 2. Mean alongshore (u) and offshore (v) velocities as functions of distance from the shore during a representative flow episode in Lake Huron [from Murthy and Dunbar, 1981].

brought to zero at the shoreline. Seaward of the flow maximum is a zone of decreasing alongshore flow predominance. Within this outer boundary layer, the midlake rotary oscillations gradually adjust to the coastal constraint. These flow patterns are entirely consistent with our earlier discussions of multiple length scales and frictional adjustment within the CBL.

The most convincing examples of significant coastally trapped waves in the CBL come from observational studies in the Great Lakes. Mortimer [1963] studied the thermal structure of Lake Michigan and found upwelling/downwelling thermocline displacements to propagate cyclonically around the southern end of the lake after relaxation of the wind stress. The observed alongshore propagation speed and cross-shore trapping scale were approximately 0.5 m/s and 5 km, respectively.

Detailed evidence of alongshore wave propagation near shore comes from the IFYGL investigations of Lake Ontario. Csanady and Scott [1974] observed rather spectacular reversals of currents and thermocline displacements near the end of several wind stress episodes. After a several-day lull in the wind stress, these reversals appeared to propagate cyclonically around the entire lake, with propagation speed and trapping scales similar to those observed by Mortimer [1963]. These data were later reexamined by Clarke [1977] and Bennett and Lindstrom [1977] within the framework of analytical and numerical forced wave theory. The studies of the waves suggested a hybrid topographic/internal Kelvin wave, as might be expected in a region characterized by a strong stratification and bottom slope.

Csanady [1976] also observed wave propagation of a different sort in Lake Ontario. In this instance a barotropic flow reversal was observed with a trapping scale of approximately 10 km. Csanady has interpreted this event as the passage of a "pure" topographic wave. Examining current records during unstratified conditions, Marmorino's [1979] statistical analyses showed further evidence of topographic waves in the lake.

## The New York Bight

Observational data from CBL/inner shelf regimes of the northeastern United States come primarily from experiments in the New York Bight. Long-term current measurements from a location 4 km offshore of Little Egg Inlet, New Jersey, have been described in detail by EG&G [1976]. A 60-day late-winter subset of the wind, current, and sea level data was studied in conjunction with a two-dimensional nonlinear numerical model by Bennet and Magnell [1979]. The results of this analysis emphasized the importance of alongshore setup and its variation. It was also concluded that while nonlinear cross-shore advection was of little importance, inclusion of alongshore advection would have significantly improved model performance.

The Long Island Inner Shelf (LINS) experiment employed an extensive current meter mooring array in an area of ridge and swale topography [Swift et al., 1973] between the Jones and Fire Island inlets. A correspondence between the orientation of the ridge and swale topography and near-bottom tidal current ellipses [May, 1979] prompted speculation that tidal currents might play a role in the formation of these large-amplitude (5 to 10 m) bedforms. This possibility was further discussed by Lavelle and Swift [1982].

Measurements from 11 current meter moorings deployed within the 11 $km^2$ LINS study area were analyzed by Han and Mayer [1981]. The average alongshore components of both wind stress and current were directed upcoast (opposite to the usual shelf mean flow in this region). Han and Mayer report that 79% of the current variance was contained in a barotropic mode without significant horizontal structure. Apparently, 40% of this variance was correlated with local wind forcing.

The Coastal Boundary Layer Transect (COBOLT) experiment, conducted off Long Island's south shore, was similar in concept to the highly successful IFYGL investigations of Lake Ontario. A principal objective of the program was to obtain detailed time series and synoptic observations of hydrographic and flow structures across the oceanic CBL. The experimental design included four current, temperature, and conductivity moorings arranged in a 12-km cross-shore transect along which synoptic shipboard measurements were also taken. The study area was located near Shinnecock Inlet some 80 km east of the LINS site and 60 km west of Long Island's eastern extremity, Montauk Point.

Some preliminary results of a COBOLT pilot experiment were presented by Scott and Csanady [1976]. More detailed analyses of the tidal, wind-driven, and externally forced flow components during the main experiment have been undertaken by

May [1979], Pettigrew [1981], Churchill [1984], and Hopkins and Swoboda [1986].

The depth-averaged currents at 3, 5, and 12 km from shore were studied in some detail by Pettigrew [1981]. Although the alongshore currents displayed an underlying trend which was uncorrelated with local atmospheric forcing, there was also a clear response of approximately 20 cm/s for 1 $dyn/cm^2$ wind stress. Results of a regression analysis suggested that the same 1-$dyn/cm^2$ wind stress produced an opposing alongshore sea surface slope of $3 \times 10^{-7}$. Wind-driven response accounted for roughly half of the subtidal current and pressure gradient variance.

One of the most striking features of the COBOLT data, and a clear contrast to the results of the LINS program, is the cross-shore structure of both alongshore and cross-shore flow components. In nearly 20% of the record, variation of the depth-averaged alongshore flow took the form of a complete flow reversal within the CBL. During these alongshore flow reversals the nearshore water column flowed downwind, while farther from shore there was transport against the wind.

This flow pattern, an example of which is shown in Figure 3, is a clear manifestation of the topographically controlled competition between wind stress and pressure gradient forces discussed earlier. A detailed consideration of these effects, including the generation of the nearshore pressure gradient field, and the predicted locus of alongshore current reversal along the coast, is presented in a model by Pettigrew [1981]. Pettigrew attributes this unique circulation pattern to the proximity of Montauk Point and Long Island Sound. The model predicts an inner region of downwind transport which broadens westward from the tip of Long Island. The current reversal is calculated to be roughly 10 km from shore at the observation site, while no reversal and little structure is predicted farther west in the LINS study area.

Another important aspect from the COBOLT data is the observation of substantial net offshore transport. Given that this offshore transport is too large to be easily dismissed as experimental error, continuity requires the coastal circulation pattern to be three dimensional in character. A consequence of such a cellular circulation pattern is that the Coriolis force associated with cross-shore transport may not be neglected. In fact, Pettigrew [1981] found this term to be one of the largest in his alongshore budget calculations.

Although a divergent offshore flow in the COBOLT region is a predicted feature of the CBL circulation model, the observed magnitude (approximately 1 cm/s at 6 km and 2 cm/s at 12 km) is too large to be attributed to the wind. Hopkins and Swoboda [1986] have suggested that a substantial portion of the observed cross-shore and alongshore flow in this region is due to external forcing by large-scale pressure fields. Churchill [1984] has shown examples of density gradients along the Long

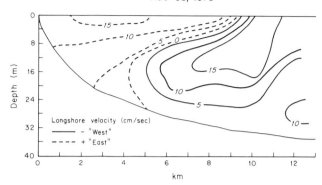

Fig. 3. Current meter transect across the CBL south of Long Island showing an alongshore current reversal following a positive alongshore wind stress event of approximately 0.5 dyn/cm² for 1.5 days [from Pettigrew, 1981].

Island coast which are of the proper sense and order of magnitude to account for the observed cross-shore transport.

## Multi-inlet Coast:
## The Southeastern United States

The coastal waters of Georgia and South Carolina, characterized by a periodic spacing of tidal inlets every 15 km, have received considerable attention in recent years. The hydrography of the first 10 km from the coast is dominated by tidal/riverine effluent plumes [Blanton and Atkinson, 1978]. The next 10 to 20 km is characterized by a complex distribution of salinity and turbidity fronts relict from preceding tidal cycles. The cross-shore salinity distributions in this intermediate zone closely resemble those of more classical coastal boundary layers but are highly variable in the alongshore direction, generally reflecting the discharge of several adjacent inlets. The seaward side of the CBL is often separated from the shelf water by a well-defined front.

Blanton [1980] described an experiment aimed at evaluating the alongshore variations unmeasured in his earlier work. Velocity and salinity data were recorded over four tidal cycles from two ships, both anchored 13 km from the coast and separated by 9 km alongshore. Despite this minimal separation, there was a 90° shift in the major axis of the tidal current ellipse, and freshwater discharge and salt flux at the two stations were unrelated. Blanton concluded that the extreme nonlinearities in the current fields, a result of the multi-inlet coast, set up a tidal residual circulation which greatly complicated material fluxes in the coastal boundary layer.

The baroclinic coastal current and the frontal zone bounding the shelf water in this region were studied by Blanton [1981]. Two ships were anchored on each side of the front (separation of 3.5

km) over four tidal cycles and time- and depth-averaged momentum-budget calculations were performed. Although no error analysis is presented, and a balance is not even approximately achieved, Blanton concludes that the Coriolis term, surface and bottom friction, and cross-shore tidal stress are all important in the cross-shore balance. Although not estimated, the baroclinic pressure gradient term is also considered of prime importance.

In the alongshore balance, Blanton [1981] reports that the Coriolis term, the alongshore slope, and the wind stress control the dynamics of the basic flow. Bottom friction and a tidal stress term are significant but of secondary importance. It is noteworthy that the data show a net offshore flow in violation of a two-dimensional mass balance and clearly illustrate the fundamentally three-dimensional nature of the flow.

Analysis of 8 years of wind and salinity data and intermittent current observations by Blanton and Atkinson [1983] shows systematic changes in behavior through the climatologic year. The geostrophic tendency for a southward flowing baroclinic coastal current is reinforced in autumn by southwestward wind stress, which pushes low-salinity water as far south as Florida. The front separating the inner shelf from the well-mixed shelf water is well defined at this time, and freshwater storage on the inner shelf is at a maximum. The onshore-directed Ekman flux in the surface layer helps maintain the front and the cross-shore density gradients, as noted in other locations by Royer [1982] and Murray and Young [1984].

During spring and summer, wind conditions reverse and slow or stall the southward flowing coastal current. Offshore Ekman flux in the surface may deplete the freshwater content of the inner shelf and weaken the front. Figure 4 presents spring/summer proportional freshwater loss rates on the inner shelf versus alongshore wind

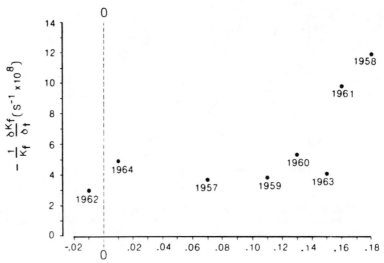

Fig. 4. The proportional loss rate of inner-shelf freshwater content $(-k_f^{-1}\partial k_f/\partial t)$ versus mean alongshore wind stress [from Blanton and Atkinson, 1983].

stress for the years 1957 to 1964. The data show a background proportional loss rate of $4 \times 10^{-8}$ $s^{-1}$, which Blanton and Atkinson [1983] attribute to tidal mixing. During years with relatively high mean wind stress values (>0.15 dyn/cm$^2$), rates increase several times as offshore Ekman transport of fresh water becomes important. Unusually strong northward wind stresses at times of high freshwater discharge have been observed to break down the front to such an extent that fresh water spills seaward, forming a two-layer shelf system.

Some preliminary work has been done in the coastal boundary layer waters north of Cape Romaine, South Carolina, an area of few inlets and insignificant freshwater discharge. Observations of Schwing et al. [1983] showed a more classical CBL, with the alongshore wind the dominant driving force of the subtidal current, and upwelling and downwelling responses alternating at a 6-day period owing to frontal passages.

## The Louisiana-Texas Inner Shelf

The massive discharges of fresh water (39,000 m$^3$/s in April) from the Mississippi and Atchafalaya rivers into the Gulf of Mexico clearly favor thermohaline circulation along the western Louisiana and Texas coasts. Using data from the Strategic Petroleum Reserve Program (SPRP), Lewis [1979a, b] has described in detail a baroclinic structure confined to within 20 to 25 km of the coast off Freeport, Texas.

The salinity distribution is clearly in phase with the discharges of the Mississippi and Atchafalaya rivers, with density gradients strongest during winter and spring months, weak during the summer, and nonexistent during the late fall. From limited current meter data, Lewis noted the presence of a baroclinic coastal jet and identi-

fied a frontal system at its seaward edge. Two distinct hydrographic/current regimes were identified. Southwestward wind stresses (downwelling favorable) were clearly associated with a fresher water (coastal jet) regime nearshore. Lewis noted that even the heavy seas of winter are generally unable to break down the frontal zone associated with the nearshore boundary layer. Summer winds (May-August) exert northeastward alongshore (upwelling favorable) wind stresses, resulting in a more homogeneous water column. In this regime the brackish water is presumably driven back toward Louisiana or offshore.

In a later report, Kelly et al. [1983] presented considerably more data on the inner-shelf currents and hydrography from SPRP sites off western Louisiana and east Texas some 200 km west (downdrift). Kelly et al. report that wind stress is the predominant motive force at the east Texas site and that northwestward prevailing winds drive westward alongshore currents during most of the year. A summer seasonal current reversal to the east closely tracks a reversal of the alongshore wind stress.

Little coherence between alongshore components of wind and currents was found at the western Louisiana site (closer to the freshwater sources) at periodicities greater than 4 days. Cross-spectrum analyses of yearlong current records suggest, however, that wind-generated motions of roughly 3-day period propagate alongshore as topographically trapped waves.

Data from the SPRP site on the inner shelf off western Louisiana were also studied by Crout [1983] with quite different results. Crout reports that during the summer, winds shift to northeastward and cause a current reversal to the east, just as observed by Lewis in east Texas. Apparently, the interplay of thermohaline and wind forcing in this region is quite variable.

Cold-front passages of several distinct types dominate the winter wind patterns. With fronts associated with extratropical cyclones, winds rotate clockwise as the front passes, and currents follow the alongshore wind shift with very little lag. Crout [1983] concludes that the alongshore momentum balance is essentially frictional, with wind stress balancing bottom stress. In spring the momentum balance is thought to include very large alongshore pressure gradients, associated with the seasonal high discharge of the Mississippi and Atchafalaya rivers.

The central Texas inner-shelf waters several hundred kilometers farther downdrift of the freshwater sources have been studied by Smith [1978, 1980b], who found baroclinic contributions to the current to be insignificant. Wind stress appears to be almost the exclusive driving force. Currents observed at an inner-shelf station compared with those made at midshelf showed higher values of kinetic energy for very low frequency alongshore motion, greater coupling with the wind stress, and less kinetic energy in the cross-shore motions at all frequencies.

## Trade Wind Zone:
## Caribbean Coast of Nicaragua

Details of coastal boundary currents along a high-rainfall coast in the trade wind zone are given by Murray et al. [1982]. Along the Caribbean coast of Nicaragua, rainfall rates of 3 to 5 m/yr produce a well-organized nearshore current with transport of approximately $5 \times 10^4$ $m^3$/s. The observed 20- to 25-km cross-shore length scale of the flow field compares favorably with the local internal Rossby radius, calculated to be roughly 25 km. The steadiness of the trade wind and the negligible tide and tidal current permitted quasi-synoptic observations of the CBL current at four locations over an alongshore distance of about 100 km. Crout and Murray [1979] report that a well-defined jet meanders southward along the coastline, in one place turning nearly perpendicular into the coast.

The dynamics of this coastal current were discussed by Murray and Young [1985]. The baroclinic and barotropic pressure gradients are the largest (and opposing) forces throughout the water column, with the Coriolis force associated with alongshore flow essentially in geostrophic balance with the net pressure gradient. Friction is important in the upper 3 to 4 m (e.g., the stratified Ekman depth), but thermohaline forcing from the freshwater input to the coast is judged to be the main driving force. Persistent trade winds consistently drive a surface Ekman layer toward the coast, thus acting to maintain the strong density gradients which characterize the CBL in this region.

## The Alaska Coastal Current

Recent research in the Gulf of Alaska has shown the coastal zone to be dominated by precipitation

and runoff effects. Royer [1979] studied the seasonal cycle of dynamic height fluctuation in coastal waters and found it to be well correlated in several locations with seasonal changes in runoff and precipitation. The majority of seasonal sea level variations were found to be the result of local steric changes. Limited data suggested that this freshwater influx drove a nearshore band of current. Satellite-tracked drifters [Royer et al., 1979] were observed trapped in a shore-parallel inner-shelf current for over 300 km.

This nearshore band of current is specifically identified as a baroclinic coastal jet (the Kenai Current) and was studied with hydrographic and current meter observations by Schumacher and Reed [1980]. Schumacher and Reed emphasized that the kinetic energy of the Kenai Current is restricted to the distinct dilute band (15 to 30 km wide) of inner-shelf water whose salinities are generally 0.5 per mil less than those of contiguous shelf water. Correlation studies showed that the baroclinic transport within this hydrographic boundary layer responds clearly to the annual cycle of freshwater influx but does not appreciably increase during the high wind speeds of winter. Discharges are indeed significant, varying from a low of about $0.3 \times 10^6$ $m^3$/s up to $1.0 \times 10^6$ $m^3$/s during October, when the integrated effect of precipitation, river discharge, and meltwater attains a maximum. The alongshore extent of the current is apparently of the order of 1000 km.

Further studies by Royer [1981] support and extend the conclusions of Schumacher and Reed [1980]. Statistical studies of the transport anomaly (departure from yearly means) of the coastal current indicate that these anomalies vary closely with the freshwater discharge to the coast and only secondarily (and poorly) with wind driving. Outside the dilute nearshore band, wind stress forcing becomes important.

Reed and Schumacher [1981] examined sea level deviations (annual mean minus monthly means) at six stations around the Gulf of Alaska and interpreted them in terms of the coastal current regimes. The regional variability of the result is considerable. The eastern Gulf is characterized by a barotropic response to wind driving in winter, while in the central Gulf the CBL/inner shelf is dominated by the thermohaline forcing of the baroclinic Kenai Current. In the western Gulf, seasonal effects are much less apparent, as is the distinction between the CBL and offshore regimes.

### Summary

Observational data from oceanic CBL/inner-shelf regions have shown a dynamical similarity to earlier observations in the North American Great Lakes. Investigations in a variety of locations have confirmed the important influences of the coastal constraint, depth variations, and horizontal and vertical density gradients. Whether each of these factors plays a dominant or subordinate role depends upon the geographic and climatological domain of the coastal zone.

Evidence has been presented which strongly suggests that coastal currents are fundamentally three dimensional in character and are dependent upon upcoast, as well as local, forcing. Wind stress is generally the principal motive force for CBL circulation, but in areas of high riverine input, thermohaline forcing can predominate. Along coasts with distributed freshwater sources (e.g., Alaska and Nicaragua), baroclinic coastal currents may maintain their identity over distances of the order of 1000 km.

Momentum-budget calculations have generally indicated cross-shore geostrophic balance. Although the alongshore balance is not easily resolved, it is apparently subject to substantial spatial and temporal variation. Important mixing and exchange processes in the coastal boundary layer remain poorly understood and difficult to measure. The observations of net cross-shore transport associated with cellular circulation patterns suggest that the exchange with offshore waters is a large-scale, rather than local, process.

Determining the influence of shelf and surf zone flow regimes upon the coastal boundary layer is an important and difficult problem. The resolution of this issue awaits comprehensive theoretical and field investigations.

Acknowledgments. This work was supported by the Office of Naval Research Coastal Sciences contracts N00014-83-K-0618 and N00014-83-C-0150. The authors are indebted to G. T. Csanady for his comments on the first draft and to Frank O. Smith for preparation and editing of the manuscript.

References

Allen, J. S., Upwelling and coastal jets in a continuously stratified ocean, J. Phys. Oceanogr., 3, 245-257, 1973.

Beardsley, R. C., and J. Hart, A simple theoretical model for the flow of an estuary onto a continental shelf, J. Geophys. Res., 83, 873-883, 1978.

Bennett, J. R., On the dynamics of wind-driven currents, J. Phys. Oceanogr., 4, 400-414, 1974.

Bennett, J. R., and E. J. Lindstrom, A simple model of the Lake Ontario coastal boundary layer, J. Phys. Oceanogr., 7, 620-625, 1977.

Bennett, J. R., and B. A. Magnell, A dynamical analysis of currents near the New Jersey coast, J. Geophys. Res., 84, 1165-1175, 1979.

Birchfield, G. E., The response of a circular model Great Lake to a suddenly imposed wind stress, J. Geophys. Res., 74, 5547-5554, 1969.

Blanton, J. O., Some characteristics of nearshore currents along the north shore of Lake Ontario, J. Phys. Oceanogr., 4, 415-424, 1974.

Blanton, J. O., The transport of fresh water off a multi-inlet coastline, in Estuarine and Wetland Processes, edited by P. Hamilton and K. B. MacDonald, pp. 49-64, Plenum, New York, 1980.

Blanton, J. O., Ocean currents along a nearshore frontal zone on the continental shelf of the southeastern United States, J. Phys. Oceanogr., 11, 1627-1637, 1981.

Blanton, J. O., and L. P. Atkinson, Physical transfer processes between Georgia tidal inlets and nearshore waters, in Estuarine Interactions, edited by M. L. Wiley, pp. 514-532, Academic, Orlando, Fla., 1978.

Blanton, J. O., and L. P. Atkinson, Transport and fate of river discharge on the continental shelf of the southeastern United States, J. Geophys. Res., 88, 4730-4738, 1983.

Bowen, A. J., The generation of longshore currents on a plane beach, J. Mar. Res., 27, 207-215, 1969.

Brink, K. H., Low-frequency free wave and wind-driven motions over a submarine bank, J. Phys. Oceanogr., 13, 103-116, 1983.

Churchill, J. H., Analysis of flow within the coastal boundary layer off Long Island, New York, WHOI Tech. Rep. 84-14, 67 pp., Woods Hole Oceanographic Institution, Woods Hole, Mass., 1984.

Clarke, A. J., Observational and numerical evidence for wind-forced coastal trapped long waves, J. Phys. Oceanogr., 7, 231-247, 1977.

Crepon, M., Hydrodynamique marine en regime impulsionnel, Cah. Oceanogr., 19, 847-880, 1967.

Crepon, M., Hydrodynamique marine en regime impulsionnel, Cah. Oceanogr., 21, 863-877, 1969.

Crout, R. L., Wind-driven, near-bottom currents over the west Louisiana inner continental shelf, Ph.D. thesis, Louisiana State Univ., Baton Rouge, 1983.

Crout, R. L., and S. P. Murray, Shelf and coastal boundary layer currents, Miskito Bank of Nicaragua, Proc. Coastal Eng. Conf. 16th, 2715-2729, 1979.

Csanady, G. T., Large-scale motion in the Great Lakes, J. Geophys. Res., 72, 4151-4161, 1967.

Csanady, G. T., Wind-driven summer simulation in the Great Lakes, J. Geophys. Res., 73, 2579-2589, 1968.

Csanady, G. T., On the equilibrium shape of the thermocline in the shore zone, J. Phys. Oceanogr., 1, 263-270, 1971.

Csanady, G. T., The coastal boundary layer in Lake Ontario, I, The spring regime, J. Phys. Oceanogr., 2, 41-53, 1972a.

Csanady, G. T., The coastal boundary layer in Lake Ontario, II, the summer-fall regime, J. Phys. Oceanogr., 2, 168-176, 1972b.

Csanady, G. T., Wind-induced barotropic motions in long lakes, J. Phys. Oceanogr., 3, 429-438, 1973.

Csanady, G. T., Topographic waves in Lake Ontario, J. Phys. Oceanogr., 6, 93-103, 1976.

Csanady, G. T., The coastal jet conceptual model in the dynamics of shallow seas, in The Sea, vol. 6, edited by E. D. Goldberg, I. N. McCave, J. J. O'Brien, and J. H. Steele, pp. 117-144, John Wiley, New York, 1977.

Csanady, G. T., The arrested topographic wave, J. Phys. Oceanogr., 8, 47-62, 1978.

Csanady, G. T., The pressure field along the western margin of the North Atlantic, J. Geophys. Res., 84, 4905-4914, 1979.

Csanady, G. T., Circulation in the Coastal Ocean, 279 pp., D. Reidel, Hingham, Mass., 1982.

Csanady, G. T., Circulation induced by river inflow in well mixed water over a sloping continental shelf, J. Phys. Oceanogr., 14, 1703-1711, 1984.

Csanady, G. T., and J. T. Scott, Baroclinic coastal jets in Lake Ontario during IFYGL, J. Phys. Oceanogr., 4, 524-641, 1974.

Curtis, E. M., Wave induced longshore currents outside the surf zone, M.S. thesis, Mass. Inst. Technol., Cambridge, 1979.

Dolata, L. F., and W. Rosenthal, Wave setup and wave-induced currents in coastal zones, J. Geophys. Res., 89, 1973-1982, 1984.

EG&G, Environmental Consultants, Summary of oceanographic observations in New Jersey coastal waters near 39°28'N latitude and 74°15'W longitude during the period May 1974 through May 1975, a report to Public Service Electric and Gas Company, Newark, N.J., 1976.

Gill, A. E., and E. H. Schumann, The generation of long shelf waves by the wind, J. Phys. Oceanogr., 4, 83-90, 1974.

Han, G. C., and D. A. Mayer, Current structure on the Long Island inner shelf, J. Geophys. Res., 86, 4205-4214, 1981.

Hopkins, T. S., and A. L. Swoboda, The nearshore circulation off Long Island, August 1978, Cont. Shelf Res., in press, 1986.

Hsueh, Y., and C. Y. Peng, A diagnostic model of continental shelf circulation, J. Geophys. Res., 83, 3033-3041, 1978.

Kao, T. W., The dynamics of fronts, II, Shelf water structure due to freshwater discharge, J. Phys. Oceanogr., 11, 1215-1223, 1981.

Kelly, F. J., J. D. Cochrane, R. E. Randall, and J. E. Schmitz, Physical oceanography, in Evaluation of Brine Disposal From the West Hackberry Site of the Strategic Petroleum Reserve Program, vol. II, pp. 1-180, Texas A&M Univ./Texas A&M Res. Fdn., College Station, 1983.

Lavelle, J. W., and D. J. P. Swift, Near-shore currents measured in ridge-and-swale topography off Long Island, New York, J. Geophys. Res., 87, 4190-4194, 1982.

Lewis, J. K., Coastal frontal systems as a pollutant control mechanism for offshore energy production, Proc. 1979 Mar. Tech. Soc. Mtg., New Orleans, pp. 389-396, 1979a.

Lewis, J. K., Physical oceanography, I, in Evalutaion of Brine Disposal From the Bryan Mound Site of the Strategic Petroleum Reserve Program, vol. I, pp. 1-94, Texas A&M Univ./Texas A&M Res. Fdn., College Station, 1979b.

Longuet-Higgins, M. S., Double Kelvin waves with continuous depth profiles, J. Fluid Mech., 34, 49-80, 1968.

Longuet-Higgins, M. S., Longshore currents generated by obliquely incident sea waves, 1, J. Geophys. Res., 75, 6678-6789, 1970a.

Longuet-Higgins, M. S., Longshore currents generated by obliquely incident sea waves, 2, J. Geophys. Res., 75, 6790-6801, 1970b.

Longuet-Higgins, M. S., and R. W. Stewart, Radiation stress and mass transport in gravity waves, J. Fluid Mech., 13, 481-504, 1962.

Malone, F. D., An analysis of current measurements in Lake Michigan, J. Geophys. Res., 73, 7065-7081, 1968.

Marmorino, G. O., Low-frequency current fluctuations in Lake Ontario, winter 1972-1973 (IFYGL), J. Geophys. Res., 84, 1206-1214, 1979.

May, P. W., Analysis and interpretation of tidal currents in the coastal boundary layer, Sc.D. thesis, 197 pp., Mass. Inst. of Technol. and the Woods Hole Oceanogr. Inst., Cambridge, 1979.

Mortimer, C. H., Frontiers in physical limnology with particular reference to long waves in rotating basins, Publ. 10, pp. 9-42, Great Lakes Res. Div., Univ. of Mich., Ann Arbor, 1963.

Murray, S. P., and M. Young, The nearshore current along a high rainfall tradewind coast - Nicaragua, Coastal Estuarine Shelf Res., in press, 1985.

Murray, S. P., S. A. Hsu, H. H. Roberts, E. H. Owens, and R. L. Crout, Physical processes and sedimentation on a broad, shallow bank, Estuarine Coastal Shelf Sci., 14, 135-157, 1982.

Murthy, C. R., and D. S. Dunbar, Structure of the flow within the coastal boundary layer of the Great Lakes, J. Phys. Oceanogr., 11, 1567-1577, 1981.

Niebauer, H. J., T. Green, and R. A. Ragotzkie, Coastal upwelling/downwelling cycles in southern Lake Superior, J. Phys. Oceanogr., 7, 918-927, 1977.

Noble, M., and B. Butman, Low-frequency wind-induced sea level oscillations along the east coast of North America, J. Geophys. Res., 84, 3227-3236, 1979.

Pedlosky, J., Longshore currents, upwelling, and bottom topography, J. Phys. Oceanogr., 4, 214-226, 1974.

Pettigrew, N. R., The dynamics and kinematics of the coastal boundary layer off Long Island, Ph.D. thesis, 262 pp., Mass. Inst. of Technol. and the Woods Hole Oceanogr. Inst., Cambridge, 1981.

Reed, R. K., and J. D. Schumacher, Sea level variations in relation to coastal flow around the Gulf of Alaska, J. Geophys. Res., 86, 6543-6546, 1981.

Rodgers, G. K., The thermal bar in Lake Ontario, spring 1965 and winter 1965-66, Proc. Conf. Great Lakes Res. 9th, 369-374, 1965.

Royer, T. C., On the effect of precipitation and runoff on coastal circulation in the Gulf of Alaska, J. Phys. Oceanogr., 9, 555-563, 1979.

Royer, T. C., Baroclinic transport in the Gulf

of Alaska, II, A fresh-water driven coastal current, J. Mar. Res., 39, 251-266, 1981.

Royer, T.C., Coastal freshwater discharge in the northeast Pacific, J. Geophys. Res., 87, 2017-2021, 1982.

Royer, T. C., D. V. Hanson, and D. J. Pashinski, Coastal flow in the northern Gulf of Alaska as observed by dynamic topography and satellite-tracked drogued drift buoys, J. Phys. Oceanogr., 9, 785-801, 1979.

Schumacher, J. D., and R. K. Reed, Coastal flow in the northwest Gulf of Alaska: The Kenai Current, J. Geophys. Res., 85, 6680-6688, 1980.

Schwing, F. B., B. Kjerfve, and J. E. Sneed, Near-shore coastal currents on the South Carolina continental shelf, J. Geophys. Res., 88, 4719-4729, 1983.

Scott, J. T., and G. T. Csanady, Nearshore currents off Long Island, J. Geophys. Res., 81, 5401-5409, 1976.

Simons, T. J., Resonant topographic response of nearshore currents to wind forcing, J. Phys. Oceanogr., 13, 512-523, 1983.

Smith, N. P., Temporal variations of nearshore inertial motion in Lake Superior, Proc. Conf. Great Lakes Res. 15th, 673-679, 1972.

Smith, N. P., Low-frequency reversals of nearshore currents in the northwestern Gulf of Mexico, Contrib. Mar. Sci., 21, 103-115, 1978.

Smith, N. P., An investigation of vertical structure in shelf circulation, J. Phys. Oceanogr., 9, 624-630, 1979.

Smith, N. P., Temporal and spatial variability in longshore motion along the Texas Gulf Coast, J. Geophys. Res., 85, 1531-1536, 1980a.

Smith, N. P., An investigation of cross-shelf variability in shelf circulation on the north-west Gulf of Mexico, Contrib. Mar. Sci., 23, 1-15, 1980b.

Swift, D. J., D. B. Duane, and T. F. McKinney, Ridge and swale topography of the Middle Atlantic Bight, North America: Secular response to the Holocene hydraulic regime, Mar. Geol., 15, 227-247, 1973.

Verber, J. L., Inertial currents in the Great Lakes, Proc. Conf. Great Lakes Res. 9th, 375-379, 1966.

Winant, C. D., Comments on "The arrested topographic wave," J. Phys. Oceanogr., 9, 1042-1043, 1979.

# ESTUARINE-SHELF INTERACTIONS

W. J. Wiseman, Jr.

Coastal Studies Institute, Louisiana State University, Baton Rouge, Louisiana 70803

Abstract. The gravitational pattern in estuaries is often perturbed, at subtidal scales, by flows resulting from other processes. Wind forcing is the most familiar of these. Subtidal estuarine flow variability appears to be ubiquitous, but no predictive framework for these circulation patterns has yet been proposed. The estuarine-shelf exchanges driven at subtidal scales result in buoyant effluent plumes, which influence shelf chemistry and biology as well as physics. The dynamics of these plumes remains a fertile area of research, principally because of a lack of knowledge concerning mixing in stratified flows.

## Introduction

Estuaries are, by definition, semienclosed coastal bodies of water, but it is becoming increasingly clear that they cannot be treated in isolation. Their dynamics and impact depend, to a large extent, on their interaction with the inner shelf.

Morphologically, they are perturbations to the large-scale coastline variability, allowing free exchange of water with the adjacent shelf. As such, they have a pronounced effect on tidal characteristics over the adjacent shelf, tending to delay the longshore propagation of the tidal wave [Munk et al., 1970]. Because estuaries are, generally, regions in which seawater is diluted by land runoff, baroclinic pressure gradients drive a net flow of light water seaward over the heavier coastal water that is intruding into the estuary along the bottom [Pritchard, 1955]. This two-layered pattern of light surface effluent and heavier, deep inflow will be referred to as the classical estuarine circulation pattern. The evolution of this conceptual flow pattern has recently been reviewed [Beardsley and Boicourt, 1981]. The estuarine effluent plumes represent a major source of interaction between the estuary and shelf. The fronts at the plume boundaries are strong convergence and mixing zones [see Simpson and James, this volume], but as the fronts dissipate, the associated pressure gradients cause local perturbations to the shelf circulation patterns [Beardsley and Winant, 1979]. Dissolved and particulate species carried within these plumes are also important to the shelf. Sediment carried by the effluent can cause major perturbations to the nearshore bathymetry and occasionally alter the nearshore circulation patterns [Murray et al., 1981]. The importance of outwelling, the export of carbon and nutrients from a marsh/estuarine system, to the shelf ecosystem is still being actively debated by ecologists, but it appears clear that progress will be made only when one fully understands and can accurately measure the transports between the shelf and the estuary [Nixon, 1980].

Many isolated estuarine plumes are of sufficient strength locally to be a dominant mode of forcing for the shelf, e.g., the Mississippi, Amazon, and Columbia river plumes. Elsewhere, though, effluent plumes from numerous smaller rivers may interact to form a region of low salinity along the coast [e.g., Blanton, 1981], the coastal boundary layer [see Pettigrew and Murray, this volume]. The inner shelf is the mixing zone for these effluents, an area in which their salinity is brought up to that of the outer shelf. Whereas the estuary proper is often thought of as the region where fresh water derived from runoff mixes with oceanic water, this mixing is rarely completed within the estuary proper. In some situations, e.g., the Amazon [Gibbs, 1970] and the mouths of the Mississippi during flood stage [Wright, 1971], essentially no mixing takes place until the effluent is outside the confines of the river mouth. Once the effluent plume has been released from the estuary, the principal dynamic balances controlling its movement and mixing with its surroundings are altered.

Estuarine-shelf exchanges are not a unidirectional process. Significant mass and momentum transports occur from the shelf to the estuary [Elliott and Wang, 1978]. Again, such exchanges are not limited to the physical characteristics of the estuary, but include biological [Garside et al., 1978] and geological [Wright et al., 1972; Wright and Sonu, 1975] as well.

In the following, I shall attempt to summarize briefly our present understanding of estuarine-shelf exchanges at the subtidal scale and our knowledge of effluent plumes.

## Subtidal Exchange Between
## Estuaries and Shelves

In his pioneering work on estuarine dynamics, Pritchard [1955] clearly identified the role of baroclinic pressure gradients within the estuary in driving the classical pattern of nontidal exchange with the coastal ocean. Although there were data collected during the next two decades that showed circulation patterns in direct contrast to Pritchard's model, these were generally treated as measurement artifacts rather than real phenomena [Carter et al., 1979]. Pritchard ignored wind stress effects in his analysis of the James River data [Pritchard, 1956], because as in his data the stress spanned the compass rose and averaged to a near-zero value. The effects of low-frequency, time-varying wind stress, though, were ignored for many years. The influence of local wind stress was suggested in data from short field studies [Pickard and Rogers, 1959] and later, in model studies [Hansen and Rattray, 1965], distinctly identified as potentially important to estuarine dynamics. Weisberg and Sturges [1976] clearly identified the importance of such forcing for the first time from extended field records. During the analysis of 39 days of current meter data from the west passage of Narragansett Bay, they estimated coherence squared values greater than 0.8 between the flow and the along-channel winds at periods of 2 to 3 days. Their data plots suggest that both unidirectional flows and sheared regimes with opposing flows in the upper and lower layers occur at these frequencies, but the unidirectional pattern is the most important. (In a similar study of the Providence River estuary, a tributary to Narrangansett Bay, the dominant subtidal response to wind forcing was found to be strongly two layered [Weisberg, 1976].) Since the transport is largely unidirectional, such a flow regime would tend to violate continuity if it were not for the peculiar geometry of the area. A connection with the east passage of the bay at the inshore end of both passages permits exchange of water between the two basins.

Similar subtidal, wind-driven exchanges are noted elsewhere, being detected primarily from tide gage records. Kjerfve [1975] suggested that Louisiana estuaries exchange water with the shelf on time scales greater than 1 day in response to Ekman convergences at the coastline driven by the alongshore wind stress. His data sets, though, are extremely short, and the statistical significance of his data is minimal at low frequencies. In a somewhat longer study, of exchanges between Corpus Christi Bay and the shelf, the transports appeared to be driven by the cross-shelf wind stress at periods of 2 to 4 days, while they were driven by the alongshelf wind stress at longer periods [Smith, 1977]. These exchanges are extremely important volumetrically; the volumes of water exchanged during meteorologically driven events are an order of magnitude larger than exchanges driven by astronomical tides. The rela-

tive importance of alongshelf wind to cross-shelf wind in driving exchange processes appears to be a function not only of the strength of the relative stress components but also of the relative water depth in the nearshore region and of frequency [Chuang and Wiseman, 1983].

Chesapeake Bay is a long, narrow coastal plain estuary whose axis runs north-south. At its mouth it opens onto the continental shelf to its east, although the thalweg is oriented more to the southeast. Through careful analysis of tide gage records from the bay [Elliott and Wang, 1978; Wang and Elliott, 1978], subtidal disturbances within a number of well-defined frequency bands have been identified. At periods of 2 to 3 days, significant exchanges with the shelf take place. These are coherent with the along-estuary wind but incoherent with the coastal sea level. The fluxes appear to be seiches within the bay with a node at the bay entrance. At periods longer than about 4 days, water levels within the bay are coherent with coastal water levels. Furthermore, events appear to propagate up the bay from the mouth. Forcing of these events by processes over the shelf thus appears probable. When the net volume flux is determined from water level records within the bay, the largest exchanges occur at periods between 4 and 10 days. The volume flux decreases rapidly at longer periods. In the 4- to 10-day band, volume exchanges are coherent with the east-west, cross-bay wind stress. Such winds apparently drive Ekman flows within the bay in the north-south direction (the direction of the bay's longitudinal axis) and also directly drive flow east-west (out of and into the bay) at the bay's mouth. It seems that the north-south wind is ineffectual at driving very low frequency exchanges because a wind that would drive water down the bay and out onto the shelf is in the same direction as one that would cause Ekman convergence at the coastline, an increased water level, and a pressure gradient that would induce flow into the bay. These two counteracting processes would tend to cancel.

A related study [Elliott and Wang, 1978] analyzed data from a yearlong current meter mooring maintained in the Potomac River estuary. The Potomac is a tributary to the Chesapeake Bay, and as such, the bay plays the role of shelf waters for the Potomac. The annual mean flow structure within the Potomac exhibits the classical estuarine circulation pattern, but the tidally averaged flow shows this pattern only 47% of the time! The remainder of the time, the patterns observed include the opposite of the classical pattern, storage or flushing (inflow or outflow at all depths), and three-layered circulation. Empirical orthogonal function analysis further shows that the variability in the exchange patterns is due not only to local forcing by the wind but also to nonlocal forcing by processes that occur in the coastal ocean (in this case, Chesapeake Bay).

As longer records have become available, seasonal variability in the subtidal exchange patterns has been identified. Using both spectrum analyses of seasonal records of net volume flux across Chesapeake Bay mouth and complex demodulation of a yearlong record of the same variable, a seasonal contrast is noted between the winter season, when the wind systems are strong and well organized, and the summer, when they are weak and poorly organized [Wang, 1979]. Using a multiple coherence analysis to separate local wind effects from shelf effects, it is seen that while shelf processes control the shelf-estuarine net fluxes at periods longer than 4 days during the winter, shelf processes dominate the fluxes only at periods longer than 16 days during the summer.

In the northwestern Gulf of Mexico the barometrically adjusted sea level exhibits strong semiannual and annual signals. The amplitude of this seasonal signal approaches that of the tidal signal. Part of this very low frequency variation is due to steric effects, part is due to local Ekman effects over the shelf, and the remainder is suspected of being caused by seasonal variations in the curl of the large-scale wind stress field [Blaha and Sturges, 1981]. Smith [1978] maintained two month-long near-bottom moorings in the ship channel connecting Corpus Christi Bay with the shelf, one deployment during a period of falling mean sea level and the other during rising mean sea level. In both cases the signal shows significant wind-driven subtidal transport. Rarely, though, is this variability sufficient to reverse the direction of the subtidal exchange, which is into the bay during rising mean sea level and out during falling mean sea level.

Less well studied are the seasonal variations in baroclinic subtidal exchanges. Obviously, seasonal river floods, which change the freshwater flow to an estuary, will consequently alter the stratification and the baroclinic patterns of exchange between the estuary and the shelf [McAllister et al., 1959; Hanson, 1965]. Another way, though, to alter the longitudinal baroclinic pressure gradients that drive the baroclinc flow within an estuary is to alter the density of the shelf water at the mouth of the estuary. This can be accomplished, among other ways, by upwelling of dense water onto the shelf or by lateral advection of water past the mouth of the estuary. The latter process occurs at the mouth of the Magothy River, an estuary tributary to Chesapeake Bay [Pritchard and Bunce, 1959]. During the spring freshet, low-salinity runoff from the Susquehanna flows southward along the western shore of the bay. Runoff to the Magothy proper is minimal. Thus the density within the Magothy reflects that of the waters that were outside the mouth of the Magothy in the immediate past. As the density front associated with the Susquehanna flood flows past the Magothy, the longitudinal pressure gradients within the estuary reverse, as does the subtidal exchange pattern with the bay. Along the Louisiana shelf, similar processes occur when the Mississippi River floods. Runoff from the Mississippi and Atchafalaya rivers flows westward along the shelf. The meteorological events that control this flood occur over the states of the northern Mississippi Valley. The meteorology that controls runoff to the smaller Louisiana estuaries is more local in nature. At times during the Mississippi flood, the waters outside the mouth of these smaller estuaries are fresher than those inside (B. Barrett, personal communication, 1973; F. Kelly, personal communication, 1983). The baroclinic pressure gradient thus reverses direction. Similar events may occur on a shorter time scale. The increased particle displacements associated with spring tides can move light water to the mouth of the estuary and alter the longitudinal pressure gradients at specific phases of the fortnightly cycle [Hayward et al., 1982]. The fortnightly cycle in stratification observed in the subestuaries of the lower Chesapeake Bay was initially thought to be due to local mixing. Increased currents during spring tides appeared to result in increased turbulent mixing and decreased stratification, thus modulating the longitudinal baroclinic pressure gradients within the estuary [Haas, 1977]. While this process now appears not to have been solely responsible for the observed patterns in the lower Chesapeake Bay, mixing is important elsewhere. The role of tidal mixing in driving the mean circulation of estuaries tributary to the Bay of Fundy was early recognized and modeled in the laboratory [Hachey, 1934]. Vertical mixing was later noted to greatly reduce the flushing time of Baltimore Harbor below what would have been expected in the absence of such mixing [Carpenter, 1960]. The stratified upper layers of the Chesapeake Bay fill the harbor and are vertically mixed within it. The resultant water mass is heavier than the surface waters at the harbor mouth and denser than the deeper waters. The resultant longitudinal pressure gradients result in a three-layered circulation with inflow at the surface and bottom and outflow at mid-depth. This pattern, which Hachey [1934] modeled in the laboratory and Carpenter [1960] observed, has since been modeled analytically as well [Hansen and Rattray, 1972]. Other numerical studies have indicated the modified circulation patterns which result from assuming that the eddy coefficients in circulation models depend directly upon the strength of the flow rather than being specified a priori [Bowden and Hamilton, 1975] including a modulation of the circulation pattern during the course of the spring-neap cycle [Godfrey, 1980].

It is clear, from work completed to the present, that significant subtidal variability in shelf-estuarine exchange processes exists at both synoptic and longer periods. It is also clear, though, that significant geographical variability exists. Those processes which are important along the Texas coast are not necessarily the dominant processes along the Washington coast. Our ability to predict, a priori, the amplitude of exchange, or even whether exchanges will be one layered or two layered, is minimal. We have, at best, begun to define the problem and describe the pheno-

mena. There is still room for much fruitful research in the near future.

## Plume Morphology

Once the light estuarine water leaves the confines of the estuary proper, it spreads and flows
as a buoyant plume. It still possesses the iner-
tia it acquired while in the estuary. It also
stands higher than the surrounding shelf waters
because of its low density, thus generating a
pressure gradient both laterally and downstream.
As the plume flows over the shelf waters, it acquires momentum from the wind and also exchanges
momentum with the shelf waters through entrain-
ment/detrainment or mixing. Furthermore, its
dynamics are influenced by the local bottom topography. A number of descriptive studies of effluent plumes indicate the relative importance of
these processes in different settings.

In small estuaries adjacent to a coastal ocean
with a sufficiently large tidal range, the direction of the shelf-estuarine exchange reverses
during the course of a tidal cycle. At the mouths
of large rivers or small rivers in flood [Garvine,
1974], the exchange may be unidirectional for many
tidal cycles, although still modulated by coastal
tides. It is principally these large unidirectional exchanges that have been the objects of
field observation programs.

The areal extent of the identifiable plume is
dependent on the rate at which fresh water is
being supplied to the estuary [Donguy et al.,
1965; Rouse and Coleman, 1976; Garvine, 1974].
Once on the shelf, the plume is often observed to
approach and attach itself to the coast rather
than to continue to flow seaward in an unbounded
fashion. Donguy et al. [1965] attributed the
tendency to track the coastline to the Coriolis
effect. A rather important exception is the Amazon outflow, which often pinches off, leaving
large boluses of low-salinity water far offshore
of its mouth [Ryther et al., 1967; Nof, 1981].
Frequently, though, the plume flows coherently,
but in a fashion other than that which would be
dictated by Coriolis effects alone. Ambient currents on both tidal [Garvine, 1974] and seasonal
scales are known to correlate with plume trajectory, a pattern discussed in detail and modeled by
Garvine [1982]. Wind stress also correlates with
variability of plume trajectory on synoptic [Rouse
and Coleman, 1976; Bowman, 1978] and seasonal
scales [Duxbury, 1965]. Whether this correlation
is due to direct momentum transfer to the plume or
to larger-scale forcing of the ambient coastal
currents, though, is not totally clear.

Buoyant expansion is known to be important to
the spreading of the plume [Bondar, 1972], but
this will vary as ambient water is mixed with or
entrained into the plume, or plume water is de-
trained. Unfortunately, our knowledge of the
associated mixing processes in these highly stratified situations is very poor. In fact, we often
do not even know the direction of mass flux across

plume boundaries, since water is entrained across
such boundaries into the region of greatest turbulent intensity [Garvine, 1979].

Finally, local shelf topography may influence
the plume's characteristics. Where the shelf is
shoal and significant mixing has already occurred
within the estuary, such that the buoyancy of the
effluent is slight, the plume may travel an appre-
ciable distance before separating from the bottom. Such may be the case for small tidal inlets. In the case of larger river mouths, the
river mouth bar frequently provides the perturbation necessary for plume detachment. In each
case, further vertical entrainment or mixing is
eliminated until the plume separates from the
bottom.

Once again, although many observations of plume
trajectories and characteristics showing significant variability are available, we have yet to
discern a unifying pattern. Many physical processes have been identified as important in different geographical settings. We are, though, unable to specify a priori the dominant dynamical
balances in any given situation. This places se-
vere limitations on our ability to solve the very
practical problem of predicting plume dynamics.

Initial attempts at modeling the effluent as a
two-dimensional jet, while possibly appropriate
for tidal inlets adjacent to shallow shelves, are
inappropriate for the buoyant effluent from an
estuary. Balancing buoyancy effects, Coriolis
deflection of the plume, and lateral momentum
diffusion [Takano, 1954] has met with some success
in explaining observations [Donguy et al.,
1965]. This model, though, does not allow for the
seaward reduction in buoyancy of the plume, a
feature that is characteristic of all estuarine
effluents. Wright and Coleman [1971] modified a
model produced by Bondar [1969] to allow for vertical entrainment of ambient seawater. They assumed no lateral entrainment because their observations on the Mississippi River effluent showed no
lateral gradients across the plume. Momentum
diffusion was ignored, and deceleration of the
plume was associated with the entrainment of momentum from below. The neglect of lateral entrainment and diffusion does not appear to be
universally justified, though [Garvine, 1974;
McClimans, 1978]. Some success has been achieved
with numerical models of effluent dynamics. Waldrop and Farmer [1974] used the full Navier-Stokes
equations to study the near-field plume of the
Mississippi River. Their greatest difficulty lay
in an inability to maintain the frontal nature of
the plume boundaries, probably because of the
assumed Fickian diffusion, which is almost certainly inappropriate. Similar time-independent
plume models have been developed for the prediction of the fate of thermal effluents from power
plants [e.g., Stolzenbach and Harleman, 1971] and
are applicable to the prediction of the trajectory
and dispersion of estuarine effluents on the
shelf. More intuitive representations of the
frontal transfer processes [e.g., Garvine, 1979;

Stronach, 1981] have resulted in reasonable agreement of model results with observations.

Many recent efforts at modeling effluents have been predictive in nature, i.e., designed to reproduce observations of plume characteristics from a particular estuary. A notable exception, which is concerned less with prediction and more with understanding, is the work of Beardsley and Hart [1978], who have produced similarity solutions for one- and two-layered source-sink flows over the shelf. Although the model does not deal with the effects of frontal boundaries of the plume, it clearly indicates the steering properties of the lower layer of the ambient shelf waters on the plume through both drag and topographic variations of the "pycnocline," which acts as a bottom for the upper layer flow. In a similar vein, O'Donnell and Garvine [1983] have numerically studied the time-dependent dynamics of a two-dimensional plume. A two-layered model was utilized which solved the long-wave equation in the body of the plume and a shock patching technique at the frontal boundary. Temporal variations of conditions at the river mouth are seen to propagate outward as internal waves, surges, or bores and affect the motion of the frontal boundary.

### Final Comments

Historically, the influence of estuarine discharge on shelf dynamics has been considered small and, except in the case of major river mouths, modeled as a uniform leakage of salinity deficit at the coastline. Renewed concern for the dynamics of the coastal boundary layer, though, has increased interest in the details of estuarine-shelf exchanges. We have only recently begun to appreciate the importance of modulations of these exchanges at time scales longer than tidal. Recent studies offer much information concerning the variability of these exchanges, but little insight into what controls the mode, amplitude, or spectral variability of this process. Similarly, while significant advances have been made in the ability to predict plume behavior in the vicinity of a given estuary, these advances have been due, in large part, to tuning of the free bulk parameters of the models. Many of these models offer little to our understanding of the underlying processes. Thus, although we now have at our disposal the results of numerous excellent descriptive studies of estuarine-shelf exchanges and their variability, considerable effort will be required to explain coherently these observations.

There are so many avenues for fruitful research that it would be presumptuous to set priorities. From a theoretical viewpoint, one major impediment to further quantitative understanding of these processes appears to be our inability to adequately describe mixing and entrainment in stratified shear flows. It would seem that improved understanding of estuarine-shelf exchange will go hand in hand with our understanding of stratified turbulent flows. On the other hand, on a more practical note, it is clear that the estuarine and shelf flow regimes are intimately linked. Yet one's interest is usually in one regime or the other, and the cost of simultaneously modeling both is generally prohibitive. It is reasonable to ask how one might, without loss of accuracy, reduce the domain which must be modeled when interest lies solely in the shelf or estuarine circulation.

Acknowledgments. The Coastal Sciences Program of the Office of Naval Research has provided continuing contractual assistance to the Coastal Studies Institute of Louisiana State University for some years, and their support is gratefully acknowledged.

### References

Beardsley, R. C., and W. C. Boicourt, On estuarine and continental shelf circulation in the Middle Atlantic Bight, in Evolution of Physical Oceanography, edited by B. A. Warren and C. Wunsch, pp. 198-233, MIT Press, Cambridge, Mass., 1981.

Beardsley, R. C., and J. Hart, A simple theoretical model for the flow of an estuary onto a continental shelf, J. Geophys. Res., 83(C2), 873-883, 1978.

Beardsley, R. C., and C. D. Winant, On the mean circulation in the Mid-Atlantic Bight, J. Phys. Oceanogr., 9(3), 612-619, 1979.

Blaha, J., and W. Sturges, Evidence for wind-forced circulation in the Gulf of Mexico, J. Mar. Res., 39(4), 711-734, 1981.

Blanton, J. O., Ocean currents along a nearshore frontal zone on the continental shelf of the southeastern United States, J. Phys. Oceanogr., 11(12), 1627-1637, 1981.

Bondar, C., Considerations théoriques sur la dispersion d'un courant liquide de densité réduite et a niveau libre, dans un bassin contenant un liquide d'une plus grande densité, in Symposium on the Hydrology of Deltas, IAHS AISH Publ. 91, pp. 246-256, Unesco, Paris, 1969.

Bondar, C., Contributie la studiul hidraulic al Iesirii la Mare Prin Gurile Dunarii, Studii de Hidrologie, vol. XXXII, 466 pp., Institul de Meteorologie si Hidrologie, Bucarest, 1972.

Bowden, K. F., and P. Hamilton, Some experiments with a numerical model of circulation and mixing in a tidal estuary, Estuarine Coastal Mar. Sci., 3(3), 281-301, 1975.

Bowman, M. J., Spreading and mixing of the Hudson River effluent into the New York Bight, in Hydrodynamics of Estuaries and Fjords, edited by J. C. J. Nihoul, pp. 373-386, Elsevier, New York, 1978.

Carpenter, J. E., The Chesapeake Bay Institute study of the Baltimore Harbor, Proc. Annu. Conf. Md-Del Water Sewerage Assn. 33rd, 62-78, 1960.

Carter, H. H., T. O. Najarian, D. W. Pritchard, and R. E. Wilson, The dynamics of motion in estuaries and other coastal water bodies, Rev. Geophys., 17(7), 1585-1590, 1979.

Chuang, W.-S., and W. J. Wiseman, Jr., Coastal sea level response to frontal passages on the Louisiana-Texas coast, J. Geophys. Res., 88(C4), 2615-2620, 1983.

Donguy, J.-R., J. Hardiville, and J.-C. LeGuen, Le parcours maritime des Eaux du Congo, Cah. Oceanogr., XVII(2), 85-97, 1965.

Duxbury, A. C., The union of the Columbia River and the Pacific Ocean--General features, in Transactions of the Joint Conference on Ocean Science and Ocean Engineering, pp. 914-922, Marine Technology Society and American Society of Limnology and Oceanography, Washington, D.C., 1965.

Elliott, A. J., and D. P. Wang, The effect of meteorological forcing on the Chesapeake Bay: The coupling between an estuarine system and its adjacent coastal waters, in Hydrodynamics of Estuaries and Fjords, edited by J. C. J. Nihoul, pp. 127-145, Elsevier, New York, 1978.

Garside, C., G. Hull, and C. S. Yentsch, Coastal source waters and their role as a nitrogen source for primary production in an estuary in Maine, in Estuarine Interactions, edited by M. L. Wiley, pp. 565-575, Academic, Orlando, Fla., 1978.

Garvine, R. W., Physical features of the Connecticut River outflow during high discharge, J. Geophys. Res., 79(6), 831-846, 1974.

Garvine, R. W., An integral hydrodynamic model of upper ocean frontal dynamics, I, Development and analysis, J. Phys. Oceanogr., 9(1), 1-18, 1979.

Garvine, R. W., A steady state model for buoyant surface plume hydrodynamics in coastal waters, Tellus, 34(3), 293-306, 1982.

Gibbs, R. J., Circulation in the Amazon River estuary and adjacent Atlantic Ocean, J. Mar. Res., 28(2), 113-123, 1970.

Godfrey, J. S., A numerical model of the James River estuary, Virginia, U. S. A., Estuarine Coastal Mar. Sci., 11(3), 295-310, 1980.

Haas, L. W., The effect of the spring-neap tidal cycle of the James, York, and Rappahannock rivers, Virginia, U. S. A., Estuarine Coastal Mar. Sci., 5(4), 485-496, 1977.

Hachey, H. B., Movements resulting from mixing of stratified water, J. Fish Res. Board Can., 1(2), 133-143, 1934.

Hansen, D. V., Currents and mixing in the Columbia River estuary, in Transactions of the Joint Conference on Ocean Science and Ocean Engineering, pp. 943-955, Marine Technology Society and American Society of Limnology and Oceanography, Washington, D.C., 1965.

Hansen, D. V., and M. Rattray, Jr., Gravitational circulation in straits and estuaries, J. Mar. Res., 23(2), 104-122, 1965.

Hansen, D. V., and M. Rattray, Jr., Estuarine circulation induced by diffusion, J. Mar. Res., 30(3), 281-294, 1972.

Hayward, D., C. S. Welch, and L. W. Haas, York River destratification: An estuary-subestuary interaction, Science, 216, 1413-1414, 1982.

Kjerfve, B., Tide and fair-weather wind effects in a bar-built Louisiana estuary, in Estuarine Research, vol. II, Geology and Engineering, edited by E. L. Cronin, pp. 47-62, Academic, Orlando, Fla., 1975.

McAllister, W. B., M. Rattray, Jr., and C. A. Barnes, The dynamics of a fjord estuary: Silver Bay, Alaska, Tech. Rep. 62, Univ. of Washington, Seattle, 1959.

McClimans, T. A., Fronts in fjords, Geophys. Astrophys. Fluid Dyn., 11, 23-34, 1978.

Munk, W., F. Snodgrass, and M. Wimbush, Tides offshore: Transition from California coastal to deep-sea waters, Geophys. Astrophys. Fluid Dyn., 1, 161-235, 1970.

Murray, S. P., J. M. Coleman, H. H. Roberts, and M. Salama, Accelerated currents and sediment transport off the Damietta Nile promontory, Nature, 293(5827), 51-54, 1981.

Nixon, S. W., Between coastal marshes and coastal waters, A review of twenty years of speculation and research on the role of salt marshes in estuarine productivity and water chemistry, in Estuarine and Wetland Processes, edited by P. Hamilton and K. B. McDonald, pp. 437-525, Plenum, New York, 1980.

Nof, D., On the dynamics of equatorial outflows with application to the Amazon's basin, J. Mar. Res., 39(1), 1-29, 1981.

O'Donnell, J., and R. W. Garvine, A time-dependent, two-layer frontal model of buoyant plume dynamics, Tellus, 35A(1), 73-80, 1983.

Pettigrew, N. R., and S. P. Murray, The coastal boundary layer and inner shelf, this volume.

Pickard, G. L., and K. Rogers, Current measurements in Knight Inlet, British Columbia, J. Fish. Res. Board Can., 16, 635-678, 1959.

Pritchard, D. W., Estuarine circulation patterns, Proc. Am. Soc. Civ. Eng., 81, 717/1-717/11, 1955.

Pritchard, D. W., The dynamic structure of a coastal plain estuary, J. Mar. Res., 15(1), 33-42, 1956.

Pritchard, D. W., and R. E. Bunce, Physical and chemical hydrography of the Magothy River, Tech. Rep. XVII, Ref. 59-2, Chesapeake Bay Inst., Johns Hopkins Univ., Baltimore, Md., 1959.

Rouse, L. J., and J. M. Coleman, Circulation observations in the Louisiana Bight using LANDSAT imagery, Remote Sens. Environ., 5, 55-66, 1976.

Ryther, J. H., D. W. Menzel, and N. Corwin, Influence of the Amazon River outflow on the ecology of the western tropical Atlantic, I, Hydrography and nutrient chemistry, J. Mar. Res., 25(1), 69-83, 1967.

Simpson, J. H., and I. D. James, Coastal and estuarine fronts, this volume.

Smith, N. P., Meteorological and tidal exchanges between Corpus Christi Bay, Texas, and the northwestern Gulf of Mexico, Estuarine Coastal Mar. Sci., 5(4), 511-520, 1977.

Smith, N. P., Long-period estuarine-shelf exchanges in response to meteorological forcing, in *Hydrodynamics of Estuaries and Fjords*, edited by J. C. J. Nihoul, pp. 147-159, Elsevier, New York, 1978.

Stolzenbach, K. D., and D. R. F. Harleman, An analytical and experimental investigation of surface discharges of heated water, *Rep. 135*, Ralph M. Parsons Lab. for Water Resources and Hydrodynamics, Mass. Inst. of Technol., Cambridge, 1971.

Stronach, J. A., The Fraser River plume, Strait of Georgia, *Ocean Manage.*, *6*, 201-221, 1981.

Takano, K., On the salinity and velocity distributions off the mouth of a river, *J. Oceanogr. Soc. Jpn.*, *10*(3), 1-7, 1954.

Waldrop, W. R., and R. C. Farmer, Three-dimensional computation of buoyant plumes, *J. Geophys. Res.*, *79*(9), 1269-1276, 1974.

Wang, D. P., Subtidal sea level variations in the Chesapeake Bay and relations to atmospheric forcing, *J. Phys. Oceanogr.*, *9*(2), 413-421, 1979.

Wang, D. P., and A. J. Elliott, Non-tidal variability in the Chesapeake Bay and Potomac River: Evidence for non-local forcing, *J. Phys. Oceanogr.*, *8*(2), 225-232, 1978.

Weisberg, R. H., The non-tidal flow in the Providence River of Narragansett Bay: stochastic approach to estuarine circulation, *J. Phys. Oceanogr.*, *6*(5), 721-734, 1976.

Weisberg, R. H., and W. Sturges, Velocity observations in the west passage of Narragansett Bay: A partially mixed estuary, *J. Phys. Oceanogr.*, *6*(3), 345-359, 1976.

Wright, L. D., Hydrography of South Pass, Mississippi River, *J. Waterw. Harbors Coastal Eng. Div. Am. Soc. Civ. Eng.*, *97*(WW3), 491-504, 1971.

Wright, L. D., and J. M. Coleman, Effluent expansion and interfacial mixing in the presence of a salt wedge, Mississippi River delta, *J. Geophys. Res.*, *76*(36), 8649-8661, 1971.

Wright, L. D., and C. J. Sonu, Processes of sediment transport and tidal delta development in a stratified tidal inlet, in *Estuarine Research*, vol. II, *Geology and Engineering*, edited by E. L. Cronin, pp. 63-76, Academic, Orlando, Fla., 1975.

Wright, L. D., C. J. Sonu, and W. V. Kielhorn, Water-mass stratification and bed form characteristics in East Pass, Destin, Florida, *Mar. Geol.*, *12*, 43-58, 1972.

# PROCESSES THAT AFFECT STRATIFICATION IN SHELF WATERS

Larry P. Atkinson and Jackson O. Blanton

Skidaway Institute of Oceanography, Savannah, Georgia 31416

Abstract. Processes that affect stratification in the South Atlantic Bight can be divided into those that create it and those that destroy it. Stratification is created by surface heating, freshwater runoff, and rain and is destroyed by evaporation, cooling, and surface and bottom stresses. We show that freshwater runoff is the primary source of buoyancy that creates stratification in the inner and middle shelf near rivers. Heating is equally important over all of the shelf. Up to 20 mW m$^{-2}$ of mixing power may be required to destroy runoff-related stratification. Wind stress can exert a maximum of about 3 mW m$^{-2}$. Another process that creates and destroys stratification is the advection of buoyancy, a term usually neglected. We show that variations in stratification caused by advective flux of buoyancy often dominate all other processes. The power required to destroy advectively created stratification may reach 3 mW m$^{-2}$.

## Introduction

Stratification of shelf waters reflects the balance between buoyant forces that result from heating and cooling, evaporation and precipitation, and runoff, and mixing forces such as wind stress at the surface and current stress at the bottom. Since most of these forces vary at a variety of frequencies from tidal to seasonal, and even geological for that matter, we can expect variations in stratification to occur at similar frequencies. In this paper we will examine seasonal variations in observed stratification and in the processes that affect stratification. Most of our discussion will be based on data from the southeastern United States continental shelf, the South Atlantic Bight (SAB) (Figure 1). The cycle of stratification in the SAB is dramatic because of the extreme seasonal and latitudinal variations that occur there and thus is a good laboratory for examining stratification processes.

In this paper we will show that while some of the processes that affect stratification are well known, some of the most important terms, such as runoff and evaporation, are difficult to parameterize, and thus a true determination of seasonal variations in buoyancy flux cannot be determined by summing terms. We will also show that the advective transport of heat into the region may at times exceed other forms of buoyancy flux.

Available oceanographic data from the SAB were used by Atkinson et al. [1983] to create charts showing mean oceanographic conditions at monthly increments. These data were used to calculate the monthly mean bulk stratification for areas of the SAB defined by 1° of latitude and the 0-, 20-, 40- and 60-m isobaths (Figure 2). The mean bulk stratification is the simple difference between near-bottom and near-surface density. During the winter, bulk stratification is low because of high wind stress, low runoff, and negative heat flux. In early spring, bulk stratification increases first in the southern SAB because of decreased winds, increased heating, and runoff. In the summer, bulk stratification increases throughout the SAB partly because of positive heat flux and decreased wind stress but also because of subsurface Gulf Stream intrusions. By October, bulk stratification generally decreases because of increased wind stress; however, an area of increased bulk stratification occurs off southeast Florida because of southward advection of fresher coastal waters.

These mean monthly distributions lead to several conjectures regarding stratification in the SAB:

1. Stratification in the spring and summer increases because of (1) increased insolation, (2) reduced winds, and (3) subsurface intrusions of denser Gulf Stream water.

2. Stratification in the fall and winter generally decreases because of (1) decreased insolation and (2) increased winds.

3. Stratification can increase locally because of the trapping of low-salinity water along the coast.

Stratification is the net result of energy inputs creating buoyancy forces plus mixing forces that destroy stratification. We now discuss the various forms of buoyant energy.

## Flux of Buoyant Energy

Buoyancy represents a force that tends to stabilize the water column. Conservation of buoyancy,

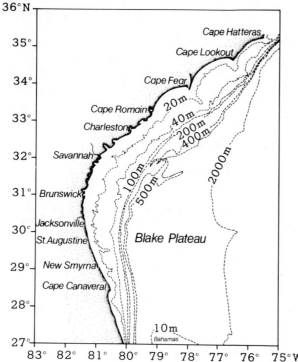

Fig. 1. The southeastern United States continental shelf area.

$$N(o) = \frac{g}{\rho}\left\{ (Q_s - LE)\frac{\alpha_v}{c_p} - ES + F \right\} \qquad (3)$$

where

| | |
|---|---|
| $N(o)$ | net flux of buoyancy across the air-sea interface; |
| $g$ | acceleration of gravity; |
| $\rho$ | density of seawater; |
| $Q(s)$ | net gain of sensible heat by radiation and conduction; |
| $L$ | latent heat of evaporation; |
| $E$ | evaporation rate; |
| $\alpha_v$ | volumetric coefficient of expansion; |
| $c_p$ | specific heat at constant pressure; |
| $S$ | salinity; |
| $F$ | freshwater flux, $kg\ m^{-2}s^{-2}$. |

Equation 3 without F is identical to Phillips' [1966, p. 223] expression.

The Q(s) term represents the rate at which buoyancy is generated by heat additions (insolation, sensible heat, convection, latent heat of evaporation). The second term, LE, is the rate at which buoyancy is lost through the latent heat of evaporation. The third term, ES, is the rate at which mass is left behind by evaporation. The fourth term, F, is the rate at which buoyancy is generated by runoff additions.

We will now quantify the terms in (3) for the SAB. In the discussion of the magnitude of buoyancy flux terms we will also give the power required to dissipate such flux distributed over a 10-m deep water column using the relation

Required Power = Buoyancy Flux $_*$ Density $_*$ Depth/2 (4)

The power is expressed in milliwatts per square meter ($mW\ m^{-2}$).

## The Heat Flux Term, Q(s)

The heat flux Q(s) (Figure 3) was determined by examining the monthly change in the mean monthly temperature in shallow (<20 m) coastal waters for the inner shelf. It was assumed that there were no advective variations in the 2° latitude section off Georgia and South Carolina over which the calculation was made. This is similar to the technique used by Pingree and Griffiths [1977] for the Celtic Sea. Using this technique, the derived quantity actually represents the sum of all heat flux terms, including latent heats. The calculated buoyancy flux varies between $-4.22 \times 10^{-8}$ and $3.28 \times 10^{-8}\ m^2\ s^{-3}$ (-0.21 to +0.16 $mW\ m^{-2}$). Maximum heat gain occurs from May through June, while maximum heat loss occurs from October through February. Insolation rates as high as $12.66 \times 10^{-8}\ m^2\ s^{-3}$ (+0.63 $mW\ m^{-2}$) have been estimated for this area from climatological data [Blanton and Atkinson, 1983]. Thus, a considerable amount of heat is probably lost by back radiation and latent heat of evaporation. We assume that these data represent the net nonadvective heat flux to all parts of the shelf.

$\beta$, can be expressed in terms of advective and nonadvective components which must balance the divergence in buoyancy flux, N, as follows [Phillips, 1966, p. 21]:

$$\frac{d\beta}{dt} = \underbrace{\frac{\partial \beta}{\partial t}}_{\substack{\text{Local}\\\text{Change}}} + \underbrace{\vec{V} \cdot \nabla\beta}_{\substack{\text{Advective}\\\text{Change}}} = \underbrace{-\nabla \cdot \vec{N}}_{\substack{\text{Mean}\\\text{Flux}}} \qquad (1)$$

where $\beta = -g(\nabla\rho/\rho_o)$ is the buoyancy per unit volume, $\vec{N} = \overline{\vec{V}'\beta'}$ is the three-dimensional buoyancy flux vector, $\vec{V}$ is the three-dimensional velocity vector, and $\nabla$ is the three-dimensional gradient operator. Primed quantities represent fluctuations about the mean. However, in the case of oceanic buoyancy flux the advective term is usually neglected, and (1) reduces to [Phillips, 1966, p. 223]

$$\frac{\partial \beta}{\partial t} = -\frac{\partial N}{\partial z} \qquad (2)$$

where $N = \overline{\omega'\beta'}$, the vertical flux of buoyancy. The implication of neglecting the advection of buoyancy will be discussed later. For the purpose of this paper we vertically integrate (2), with the left-hand side of (2) becoming the change in water column buoyancy and the right-hand side becoming the buoyancy flux across the air-sea interface as follows:

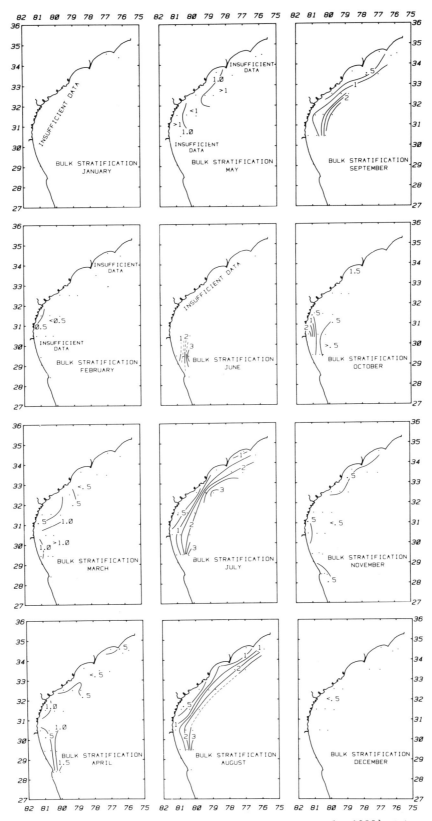

Fig. 2. Mean monthly bulk stratification [from Atkinson et al., 1983]. Units are $\sigma_t$.

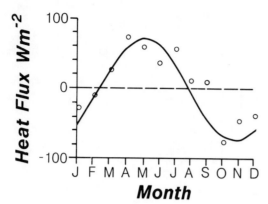

Fig. 3. Mean monthly heat flux.

## The Evaporation-Precipitation Term, ES

Evaporation in the SAB is approximately 300 cm $yr^{-1}$ [Bunker, 1976], and we must assume that it is aseasonal. Precipitation at coastal stations averages 120 cm $yr^{-1}$. It is probable that precipitation over shelf waters is less than at land stations, but data are not available to confirm this. We assume that E - P is 180 cm $yr^{-1}$ and aseasonal. This converts to a buoyancy flux of $-6.66 \times 10^{-8}$ $m^2$ $s^{-3}$ (-0.3 mW $m^{-2}$).

## The Freshwater Term, F

Because of the many small rivers, freshwater flowing into the SAB approximates a line source between Jacksonville, Florida, and Cape Fear, North Carolina. Discharge peaks in spring, and maximum monthly values based on 20-year means range between 1300 and 4000 $m^3$ $s^{-1}$ [Blanton and Atkinson, 1983] over the 400-km length of shoreline (Figure 4). The effect of coastal runoff on buoyancy flux is difficult to evaluate because unlike heat flux and evaporation, which generally occur equally over a large area, runoff is a coastal source and the manner in which it spreads over the shelf determines the resultant buoyancy flux. Blanton and Atkinson [1983] assume that the rivers flow in along 400 km of coastline and spread over the inner 20 km of the shelf. With this assumption, we calculate a buoyancy flux that ranges between 160 and 400 $\times$ $10^{-8}$ $m^2$ $s^{-3}$ (8 to 20 mW $m^{-2}$). Clearly, the buoyancy flux due to river discharge is highly variable and no doubt is the most important contribution to buoyancy flux for the inner shelf, with diminishing impact offshore depending on the degree of spreading.

### Flux of Mixing Energy (Mixing Power)

Several processes apply power at the surface and bottom boundaries which acts, along with a negative buoyancy flux, to weaken, if not destroy, stratification. Wind stress during storms can easily produce vertically mixed water on most areas of shallow continental shelves. Wind stress

also produces subtidal currents which exert stress at the bottom boundary where additional work is performed.

Tidal currents also perform work at the bottom boundary which can vertically mix stratified water. Tidal currents are of relatively constant amplitude and are at a higher frequency than wind-induced currents. Tidal power available for mixing amounts to 0.03-0.2 mW $m^{-2}$ [Blanton and Atkinson, 1983].

## Wind and Bottom Stress Work

The wind regime in the SAB is quite variable, with southerly winds occurring during the summer months and northerly winds dominating the winter months [Weber and Blanton, 1980]. The effect of wind on stability is twofold: Direct local forcing causes vertical mixing, while wind-induced advection may cause low-salinity coastal water to move into other regions, effectively causing a buoyancy flux in the offshore region or vice versa. Vertical mixing due to local wind forcing is not intense because of the relatively weak mean winds, although occasional storms may cause strong mixing. Wind-induced advection can have a substantial effect on stratification, since significant amounts of fresh water can be advected offshore [Blanton and Atkinson, 1983]. This process is particularly effective in the spring when southerly winds cause the offshore advection of fresh coastal waters.

Work that mixes the water column due to wind can be derived by multiplying wind stress by $U_*$ (friction velocity):

$$\rho_w U_*^2 = \rho_a C_d U_a^2 \qquad (5)$$

where $\rho_a$ and $\rho_w$ are the densities for air and water, respectively, $U_a$ is wind speed, $U_*$ is friction velocity, and $C_d$ is a drag coefficient. Since $\rho_a / \rho_w = 10^{-3}$, the friction velocity is

$$U_* = (10^{-3} C_d)^{1/2} U_a \qquad (6)$$

Thus the work done by the wind, $W_w$, is

$$W_w = \rho_w U_*^3 = \rho_w [(\rho_a / \rho_w) C_d]^{3/2} U_a^3 \qquad (7)$$

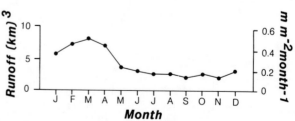

Fig. 4. Mean monthly river flow in units of total cubic kilometers and meters per square meter of shelf area.

TABLE 1a. Typical Values of Buoyancy Flux and Mixing Powers in the SAB

| Term | Typical Range | |
| --- | --- | --- |
| | Buoyancy Flux, $10^{-8}$ m$^2$ s$^{-3}$ | Equivalent Mixing Power, mW m$^{-2}$ |
| 1. Heat additions | -4.22 to 3.28 | 0 to 0.16 |
| 2. Evaporation of water | -6.66 | none required |
| 3. Runoff of fresh water | 160-400 | 8-20 |
| 4. Advective flux | 40-400 | 2-20 |
| 5. Wind stress | | 1.0-3.1 |
| 6. Bottom currents | | 0.5-1.1 |
| 7. Tidal current stress | | 0.03-0.2 |

Buoyancy flux values have been converted to power units for comparison with the mixing power data by assuming a 10-m deep water column.

The drag coefficient, $C_d$, is typically $2 \times 10^{-3}$. Mean winds in the SAB would result in $W_w = 1.0$ mW m$^{-2}$, while typical storm winds would yield a higher value, $W_w = 3.1$ mW m$^{-2}$. Similarly, the work done by bottom currents, $W_c$, is

$$W_c = \rho_w \, C_d^{3/2} U_c^3 \qquad (8)$$

where $U_c$ is the observed near-bottom current. For a typical range of bottom currents the work would be 0.5 to 1.0 mW m$^{-2}$.

Discussion of Buoyancy Flux and
Available Mixing Work

Now that we have discussed each of the normal terms in the buoyancy flux equation and the mixing work terms, it will be useful to compare them.

Table 1 shows that buoyancy flux related to freshwater runoff is an order of magnitude higher than other buoyancy sources. This buoyancy flux is maximum near the coast and decreases offshore and toward the southern and northern portions of the SAB, where runoff decreases. Thus in regions remote from runoff we would predict a net buoyancy flux closer to the sum of the heat flux and ES component. Since, as will be shown, we do observe stratification during the summertime, when buoyancy flux due to runoff is relatively low, we can assume that buoyancy flux due to the heating term Q(s) is greater than the negative buoyancy flux caused by mass left behind during evaporation (ES).

When one compares the available mixing energy to the components of buoyancy flux, it is apparent that most winds would be sufficient to dissipate any stratification resulting from a positive buoyancy flux unless runoff is present or winds are weak. The maximum possible buoyancy flux related to runoff would require more mixing work than is available during most wind events. Thus we would predict that inner shelf waters would remain stratified until strong winds during a large storm event occur or until the advection of surface waters creates a local negative buoyancy flux.

Another source of buoyancy flux is the onshore flow of dense subsurface intrusions [Atkinson, 1977]. These would create local areas of high stratification that also would require more mixing power than is available. The fact that outer shelf waters remain stratified during the summer suggests that advective buoyancy flux overcomes any local mixing forces.

Examples of the Formation and Destruction
of Stratification, April 1980

During a large experiment conducted in April 1980, when the shelf was strongly stratified in the vertical due to runoff, a storm occurred that caused stratification to break down over various parts of the shelf in response to variations in wind-induced vertical mixing. Changes in stratification were determined by calculating the potential energy of the water column relative to a

TABLE 1b. Power Required to Vertically Mix Observed Stratified Regimes

| Regime | Power, mW m$^{-2}$ |
| --- | --- |
| Inner shelf, high runoff (spring 1980) | 1.10 |
| Inner shelf, low runoff (autumn) | 0.11 |
| Inner shelf, solar heating during May | 0.02 |
| Inner and middle shelf in April 1980 | 0.0-4.00 |

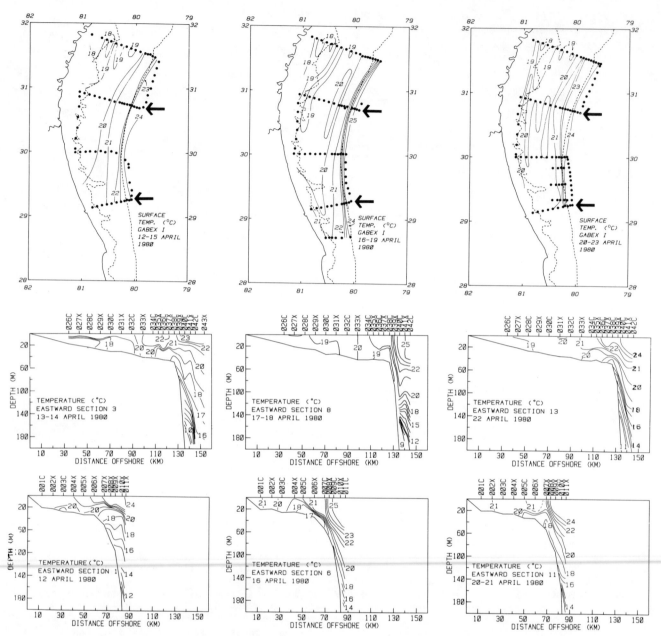

Fig. 5a. Surface and vertical temperature distributions for April 12-15, 16-19, and 20-23, 1983. Arrows indicate locations of upper and lower sections.

completely mixed state [Simpson et al., 1978]. The potential energy, PE, is calculated by

$$PE = \int_{-h}^{0} \beta z \, dz = \frac{1}{\rho} \int_{-h}^{0} (\rho - \bar{\rho}) gz \, dz \qquad (9)$$

where z is positive upward, h is the water depth, and $\rho$ is density, whose vertical average is $\bar{\rho}$.

We mapped the continental shelf off Georgia and Florida on three occasions during April 12-25, 1980. Winds were northward prior to this period and until a cold front passed on April 14. After-ward wind stress shifted southward until April 23, when stress again shifted northward. Available evidence [Lee and Brooks, 1979] suggests that the response of shelf currents to wind follows a "frictional equilibrium" model. The sequence of wind stress described above can be summarized as follows:

Before April 10: Winds weak.

April 10-17: Wind stress northward and average currents northward with offshore/onshore flow in the surface/bottom layer.

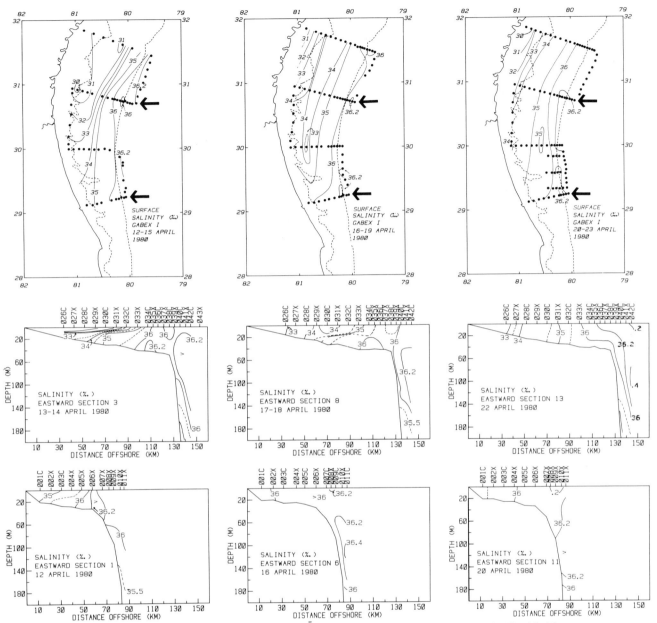

Fig. 5b. Surface and vertical salinity distributions for April 12-15, 16-19, and 20-23, 1983. Arrows indicate location of upper and lower sections.

April 17-23: Wind stress strong and southward accompanied by southward currents except for on-shore/offshore flow in surface/bottom boundary layer.

April 23-27: Wind stress and currents again northward as in the period April 10-15.

Thus the wind regime and the hydrographic response can be divided into three episodes. Episode 1 was composed of alongshelf wind stress northward from April 12 to 17; episode 2 was composed of alongshelf stress southward from April 17

to 22; and episode 3 represents a return to northward flow. We have a complete map of the hydrographic structure of the inner and middle shelf representative of each episode (Figure 5).

At the beginning of episode 1 the water was stratified at all stations. After April 17 the inner shelf was well mixed while the middle shelf continued to lose PE. Thus northward wind stress was quite effective in mixing middle shelf waters. During episode 2 the inner shelf remained well mixed while PE on the middle shelf continued

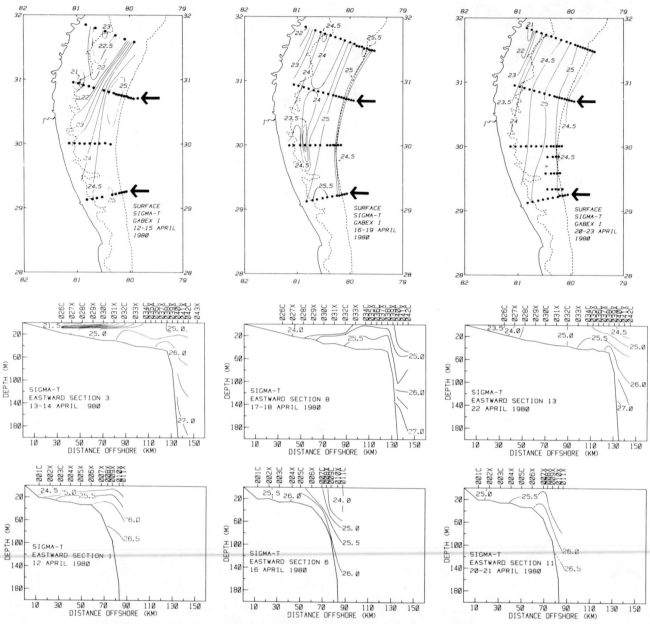

Fig. 5c. Surface and vertical density distributions for April 12-15, 16-19, and 20-23, 1983. Arrows indicate location of upper and lower sections.

to decrease at the same rate. By episode 3 the entire inner and middle shelf was nearly totally mixed.

Using equation (9), we calculated PE for the three surveys and then calculated the difference in PE at each station in each successive survey (Figure 6). Potential energy during the first mapping varied from 200 J m$^{-2}$ in inshore waters to 3000 J m$^{-2}$ over the middle shelf. During the second mapping, PE over the inner shelf was near zero, and middle shelf PE had decreased by 500 to 1500 J m$^{-2}$. By the third mapping, PE over the

middle shelf decreased another 500-1500 J m$^{-2}$. Parts of the inner and middle shelf were already vertically mixed and thus could lose no more energy.

Both middle and inner shelf decreased in PE by about 1100 J m$^{-2}$ over the 5-day period of episode 1. This represents a power of $\delta PE/\delta t = 2.25$ mW m$^{-2}$. During episode 2 the middle shelf lost about another 1000 J m$^{-2}$. The inner shelf was already well mixed, although in some places PE increased during episode 2.

Boundary work from wind and bottom stress must

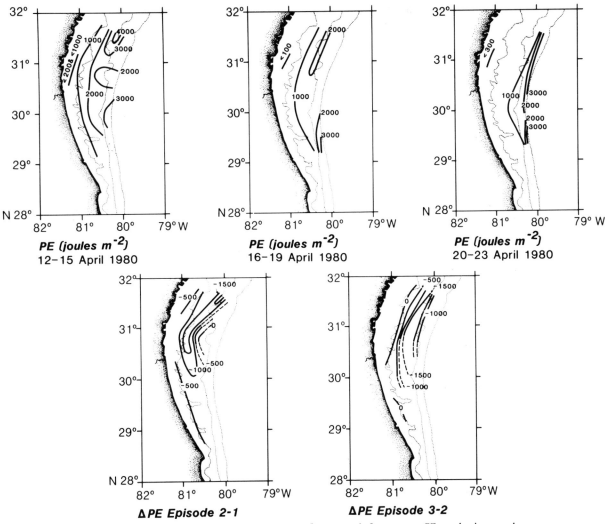

Fig. 6. Horizontal distribution of potential energy PE and change in
potential energy, ΔPE .

be approximately 2-4 mW m$^{-2}$, based on changes
suggested in Figure 6, to vertically mix the stratified waters observed on the middle and inner
shelf. We used wind speed data in equation (7)
and current speed data in equation (8) to estimate
boundary work for April 1980.

Wind data from the Savannah Navigational Light
Tower (SNLT) were used to estimate the average
work done by wind on the inner shelf. Data from a
National Oceanic and Atmospheric Administration
(NOAA) buoy on the middle shelf was used to estimate wind work there. Currents were also measured
8 m off bottom at SNLT. There was a current meter
located 3 m off bottom in 28 m of water off Savannah. The drag coefficients used for the estimates
of work were as follows:

$$C_d = 2 \times 10^{-2} \text{ (wind drag)}$$
$$C_d = 5 \times 10^{-3} \text{ (current drag at 8 m)}$$
$$C_d = 10 \times 10^{-3} \text{ (current drag at 3 m)}$$

Drag coefficients are often observed to be higher
for observations located closer to bottom [Winant
and Beardsley, 1979]. Using the above values for
$C_d$, we were able to estimate the total work done
by wind stress on the inner shelf and the total
work done by bottom currents for the first two
episodes (Table 2).

Total boundary work (Table 2) varied between
1.4 and 4.2 mW m$^{-2}$. The higher value occurred
because of the high winds and bottom currents
observed during the first episode. In all cases,
wind stress work was higher than bottom stress
work by a factor of 2 to 3. We were unable to
balance the potential energy changes observed in
the hydrography: less PE was lost than energy put
in. By combining all hydrographic stations at
depths less than 25 m in the inner shelf regime,
we observed PE changes during episode 1 that were
within 40% of the boundary work expended. PE
changes in the middle shelf regime (25-40 m) were

TABLE 2. Summary of Potential Energy Changes Compared With Work Done by Wind and Bottom Currents on the Inner and Middle Shelf for the Two Episodes

| | Inner Shelf | Middle Shelf |
|---|---|---|
| **Episode 1** | | |
| Energy lost | 1100 | 1100 |
| Wind stress work | 1340(3.10) | 690(1.60) |
| Bottom stress work | 480(1.11) | 260(0.60) |
| Total boundary work | 1820(4.21) | 950(2.20) |
| **Episode 2** | | |
| Energy lost | already mixed or increased | 1000 |
| Wind stress work | 520(1.20) | 410(0.95) |
| Bottom stress work | 260(0.60) | 200(0.46) |
| Total boundary work | 780(1.80) | 610(1.41) |

All energy values are in joules per square meter. Numbers in parentheses represent power expenditures in $10^{-3}$ mW m$^{-2}$.

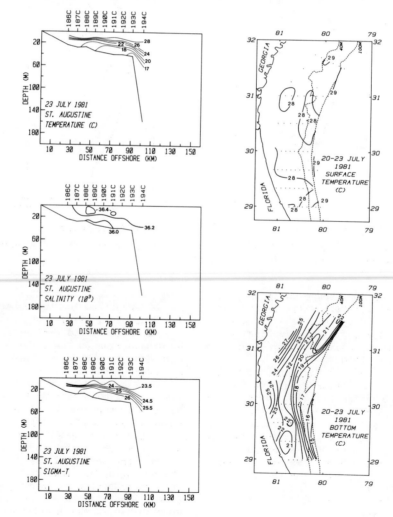

Fig. 7. Typical summertime distribution of temperature, salinity, and density.

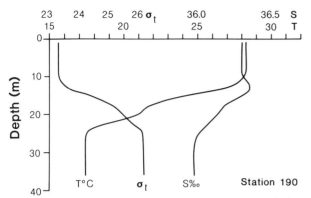

Fig. 8. Typical temperature, salinity, and density profiles.

within 20% of the available boundary work expended. These discrepancies are partially due to uncertainties in drag coefficients and the parameterizations leading to equations (7) and (8). We have neglected the advection of buoyancy between the middle and inner shelf within each episode, which would require additional boundary work to vertically mix the additional input of PE.

As an additional test on the balance we calculate the drag coefficient by assuming that the observed wind stress exactly caused the observed loss in PE. Thus, from equation (7),

$$\rho_w [(\rho_a/\rho_w)C_d]^{3/2} \, U_a^3 = \frac{\delta PE}{\delta t} \quad (10)$$

The average wind speed off Savannah during episode 1 was $U_a$ = 9.65 m s$^{-1}$. With $\rho_a/\rho_w$ = 10$^{-3}$ we calculate a drag coefficient, $C_d$, of 2.0 x 10$^{-3}$. This value, perhaps fortuitously, is identical to the "traditional" $C_d$ of 2 x 10$^{-3}$ used in many studies [Blanton, 1981].

This example demonstrates that winds, even relatively strong winds, were not sufficient to totally mix the middle shelf waters because of the excess buoyancy caused by low-salinity surface waters. Only when the surface waters were advected away did total mixing occur.

The observations we have just discussed demonstrate that stratification can be created and destroyed in a relatively short period of time depending on the balance of advective buoyancy flux and wind and bottom stress. The discussion of the observations also introduced the magnitude of the mixing processes that destroy stratification which is developed during seasonal heating and the critical nature of the balance between advective buoyancy flux and wind-related mixing.

Example of Advective Buoyancy Flux

In this paper we have so far neglected advective buoyancy flux because it is usually difficult to quantify. In this section we will use data from the summer of 1981 to show the importance of advective buoyancy flux due to intrusions.

During the summer in the SAB the total heat flux can reach 0.6 mW m$^{-2}$ and runoff is low; thus the total nonadvective buoyancy flux is in the 0.4-0.8 mW m$^{-2}$ range. Note that we are using the required mixing power for a 10-m deep water column, so this flux range would be less by a factor of 4 at the 40-m isobath. Mixing power during the summer is low, with up to 0.3 mW m$^{-2}$ from tides and less than 1 mW m$^{-2}$ from winds. However, even though the mixing power is low, it is probably higher than the nonadvective buoyancy flux. Thus we would expect to see the SAB shelf in the summer to be mixed; however, this is not the case. Evidently, because of Gulf Stream intrusions, the shelf receives a high enough advective buoyancy flux, in terms of the subsurface onshore flow of high-density water, to offset mixing forces. A typical cross-shelf section (Figure 7) shows a strong two-layered system with a very strong pycnocline. The pycnocline extends onshore but eventually breaks down, and very nearshore waters are typically mixed. The horizontal maps of surface and bottom temperatures show the extensive area impacted by the intrusions. In the absence of intrusions the bottom temperatures would be nearly the same as surface temperatures.

Vertical gradients (Figure 8) are extremely high. We have found that cold, dense Gulf Stream water invades the shelf waters periodically, producing a varying input of buoyancy. As an example of these variations we show the time variation in temperature, salinity, density (Figure 9), and PE (Figure 10) at three locations on the shelf at the 20-, 32-, and 40-m isobaths at 30°N. The surface currents at the 45-m isobath are also shown (Figure 11). The locations were highly stratified mainly because of temperature. During the study we identified three cooling events at the 45-m station but only one at the 32-m isobath, indicating the varying distance the subsurface intrusions move onshore. Examination of the current meter record suggests that the events at the 45-m isobath are related to Gulf Stream variations. At the 32-m isobath, only the event of July 25-30 was observed, reflecting the effect of the large eddy event that occurred then. The change in PE at the three stations shows that the mean at the 45-m isobath was higher than the mean at 32 or 20 m, reflecting the proximity of the Gulf Stream. Adjusting for depth, the 45-m PE is still higher than the 32- or 20-m station.

Potential energy varies by about 800, 1000, and 2000 J m$^{-2}$ over a 10-day period at the 20-, 32-, and 45-m isobaths, respectively. The variation at the 45-m isobath between July 20 and 27 was related to the passage of an eddy (see Figure 11) that caused eddy-induced upwelling at the shelf break, weakening the pycnocline and reducing the potential energy. Simultaneously, cold water was forced further onshore (see Figure 9), increasing PE at the 32-m isobath. At the 20-m isobath the onshore buoyancy flux was not observed; either it did not extend that far, or mixing removed it. PE variations at the shelf break (45 m) are more

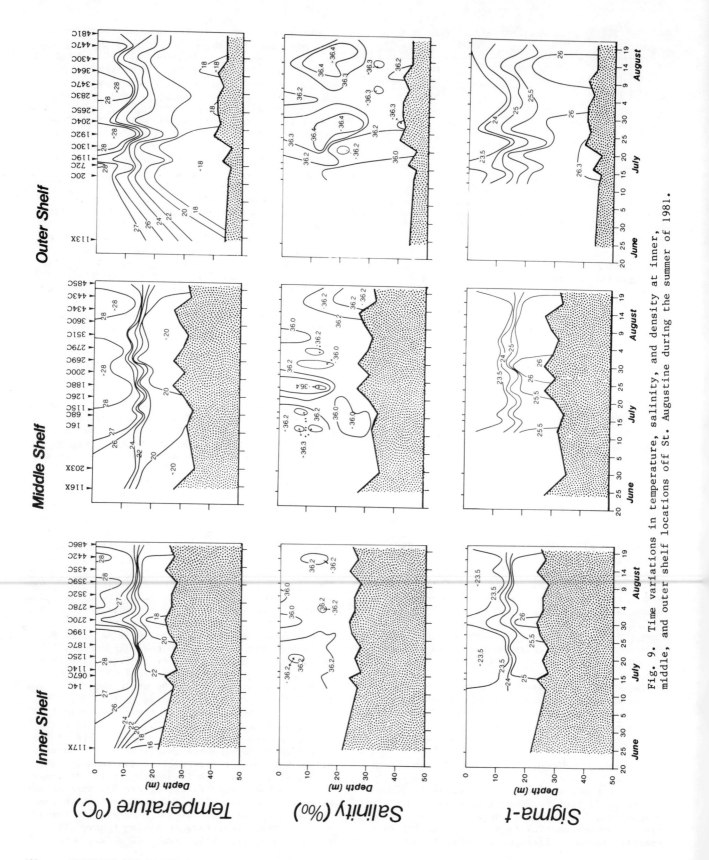

Fig. 9. Time variations in temperature, salinity, and density at inner, middle, and outer shelf locations off St. Augustine during the summer of 1981.

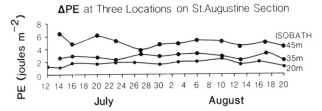

Fig. 10. Time variations in the stratification parameter (potential energy) at the three locations shown in Figure 9.

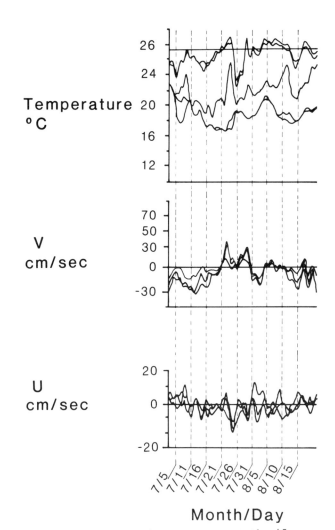

Fig. 11. Currents and temperature at the 45-m isobath. Forty-hour low-pass-filtered data from four depths are shown. Data were provided by T. Lee, University of Miami.

dependent on direct contact with the Gulf Stream, while further onshore the variations are of a more indirect nature. After the July 20-27 event, PE again increased at the 45-m isobath because of the influx of warmer surface Gulf Stream waters. At the 35-m isobath, PE remained high until wind events in mid-August caused mixing, thereby reducing PE.

During the July 20-27 intrusion event, PE increased at the 35-m isobath by about 1000 J m$^{-2}$, which would require about 1.5 mW m$^{-2}$ of mixing power to remove. This is more than could be supplied by tidal or wind-generated currents, and winds would have to be higher than normal to dissipate this buoyancy. Thus, as expected, the buoyancy advected in on July 20-27 remained.

The buoyancy increases at the outer shelf are no doubt mainly removed by advective processes rather than mixing forces, since over 2 mW m$^{-2}$ of mixing work would be required, and such energy is not available during the summer.

## Conclusions

As we have shown in the previous discussions, stratification in the SAB varies depending mainly on the balance between buoyancy provided by runoff, heating, and positive advective buoyancy flux, and mixing resulting from winds and negative advective buoyancy flux. Runoff is most important in the inner and middle shelf, while negative and positive advective buoyancy flux is more important over the middle and outer shelf. Winds are important sources of mixing energy only during storm events. Mean winds are too weak to affect stratification.

In a seasonal sense, runoff is more important during the spring, when it reaches a maximum. Advective buoyancy flux is more important when shelf waters are of a differing density than near-bottom Gulf Stream waters at the shelf break. This occurs mainly during the summer, when shelf waters are heated and subsurface intrusions are observed. Latitudinally, the effects of runoff are most pronounced adjacent to the rivers with less effect to the north and south.

Stratification variations due to advective buoyancy flux probably occur more frequently toward the south, where the density contrast is greater [Atkinson, 1977; Atkinson et al., 1983],

but can extend across the shelf at probably any location given favorable conditions. As our example has demonstrated, the stratification variations due to advective buoyancy flux can be equal to or greater than all other forms of buoyancy or mixing.

Acknowledgments. This work was supported by Department of Energy contracts DE-AS-76EV0089 to Larry P. Atkinson and DE-AS09-80EV10331 to Jackson O. Blanton with additional funding from Minerals Management Service (contract 15-830137-66). We would like to thank Bill Chandler, Pat O'Malley, and Jim Singer for their technical assistance. We thank our colleagues David Menzel, Thomas Lee, and others for helpful criticism. We are also indebted to an anonymous reviewer who corrected some mathematical inconsistencies.

## References

Atkinson, L. P., Modes of Gulf Stream intrusion into the South Atlantic Bight shelf waters, Geophys. Res. Lett., 4(12), 583-586, 1977.

Atkinson, L. P., T. N. Lee, J. O. Blanton, and W. S. Chandler, Climatology of the southeastern United States continental shelf waters, J. Geophys. Res., 88(C8), 4705-7418, 1983.

Blanton, J. O., Ocean currents along a nearshore frontal zone on the continental shelf of the southeastern U.S., J. Phys. Oceanogr., 11(12), 1627-1637, 1981.

Blanton, J. O., and L. P. Atkinson, Transport and fate of river discharge on the continental shelf of the southeastern United States, J. Geophys. Res., 88(C8), 4730-4738, 1983.

Bunker, A. F., Computations of surface energy flux and annual air-sea interaction cycles of the North Atlantic Ocean, Mon. Weather Rev., 104, 1122-1140, 1976.

Lee, T. N., and D. A. Brooks, Initial observations of current, temperature, and coastal sea level response to atmospheric and Gulf Stream forcing on the Georgia shelf, Geophys. Res. Lett., 6, 321-324, 1979.

Phillips, O. M., The Dynamics of the Upper Ocean, 261 pp., Cambridge University Press, New York, 1966.

Pingree, R. D., and D. K. Griffiths, The bottom mixed layer on the continental shelf, Estuarine Coastal Mar. Sci., 5, 399-413, 1977.

Simpson, J. H., C. M. Allen, and N. C. G. Morris, Fronts on the continental shelf, J. Geophys. Res., 83(C9), 4607-4614, 1978.

Weber, A. H., and J. O. Blanton, Monthly mean wind fields for the South Atlantic Bight, J. Phys. Oceanogr., 10, 1256-1263, 1980.

Winant, C. D., and R. C. Beardsley, A comparison of some shallow wind-driven currents, J. Phys. Oceanogr., 9(1), 218-220, 1979.